A
Review
of
Biostatistics
A
Program
for
Self-

A Review of Biostatistics

A Program for Self-Instruction

Fourth Edition

Paul E. Leaverton, Ph.D.

Professor and Chairman,
Epidemiology and Biostatistics,
College of Public Health,
University of South Florida,
Tampa, Florida

Little, Brown
and
Company
Boston/
Toronto/
London

Contents

Preface

Statistical methods are generally accepted among scientists as the most satisfactory approach yet developed for dealing with variation. In the past few decades there has been increasing recognition that variation among similarly categorized human beings "treated alike" or "exposed alike" is usually large. Consequently, throughout the biomedical and clinical sciences there has been a concomitant increase in the use of statistical methods.

While many excellent biostatistics textbooks have appeared detailing appropriate techniques, an appreciation of the subtle and not-so-subtle interpretations of calculations has not always been conveyed. *A Review of Biostatistics* (4th ed.) addresses this issue. It is not a cookbook and provides only a few formulas and examples; the emphasis is placed, instead, on basic description and interpretation of the most commonly used methods. Most readers will not become heavy users of statistical methods; if that is their goal then several of the referenced books would be quite suitable. However, all medical and health practitioners have an obligation to read their scientific literature and consequently will need to be able to understand and interpret the statistical techniques so often employed there.

This programmed syllabus reviews fundamental statistical procedures and concepts that are appropriate for those training for a career in medicine, public health, and related health professions. Most of the material relates to subjects that are necessary in critically reading and evaluating the technical, scientific literature in health and medicine.

The first three chapters include elementary coverage of definitions, rates, descriptive statistics, and population-sample concepts. Chapter 4 contains material on confidence interval interpretation and Chapter 5 emphasizes the precise interpretation of the statistical significance tests that are so often used in evaluating clinical and research data. Chapter 6 reviews linear regression and correlation. Chapters 7 and 8 present a few basic principles of clinical trials and epidemiology, because statistical methods are at the core of those two fields.

A list of objectives for this manual is provided on page ix. The objectives are written in terms of the specific behavior one might expect from a reader who has completed the program. Such a list has proved useful both for students and for instructors who use the manual with or without collaborative material.

Since the publication of the third edition in 1986, translations of this book in Greek, Japanese, Indonesian, Spanish, and Italian have appeared. It is clear that the conciseness of the material has been a key to this international popularity. Therefore, only a small amount of material has been added in this edition: There is newly added information in the chapters on descriptive statistics, probability, and epidemiology and a few clarifications throughout. I hope, with these changes, the book continues to meet its stated objectives.

P.E.L.

Educational Objectives

After completing this syllabus, the reader should be able to

I. Recognize the utility of specified descriptive statistical measures in medicine and public health.
   A. Fundamental concepts.
      1. Differentiate between discrete and continuous data.
      2. Differentiate between nominal and ordinal variables.
      3. Define a population from a statistical point of view.
   B. Population rates.
      1. Define disease prevalence rate and incidence rate, and, given the relevant population and disease data for a period of time, be able to calculate a specified prevalence and incidence rate.
      2. Define sensitivity and specificity with respect to a diagnostic test, and, given the results of such a test applied to known diseased and well persons, be able to calculate test sensitivity and specificity.
      3. Use and understand positive and negative predictive values in terms of clinical and screening tests.
   C. Descriptive statistics.
      1. Given a set of biologic measurements, construct and interpret a frequency distribution, relative frequency distribution, cumulative frequency distribution, and histogram.
      2. Given a set of biologic measurements, state the meaning of the range, variance, standard deviation, coefficient of variation, and percentiles with respect to these measurements.
      3. Given a set of biologic measurements, calculate a mean, median, and mode,

and state the relative advantages of these measures of location.

II. Understand basic concepts of probability.
   A. Define the probability of an event. Given a probability statement concerning some specific medical phenomenon, specify the relative frequency interpretation of probability in this context.
   B. Define mutually exclusive events.
   C. Define independent events.
   D. Calculate and define conditional probability given relevant basic 2×2 table categorical data summaries.

III. Know the concept of a population distribution and sampling.
   A. Describe the concept of a population distribution.
   B. Given a scientific article from a medical journal, describe the population to which the investigators intend to refer their findings.
   C. Define a random sample.
   D. Know that a set of experimental data may be regarded as a sample from a hypothetical population.
   E. Given a scientific article from a medical journal, describe how the sample was selected from the population. Delineate possible sources of bias that would make the sample nonrepresentative.
   F. Recognize a distribution that is approximately Gaussian (normal).
   G. State the kinds of data for which a standard deviation is an appropriate measure of variation.
   H. State the approximate percentage of population members included by $\mu \pm 1\sigma$, $\mu \pm 2\sigma$, and $\mu \pm 3\sigma$ if the distribution is Gaussian.
   I. Interpret the meaning of the phrase "normal limits," and, given a biologic measurement and the corresponding normal

limits, interpret this value in terms of the appropriate distribution percentiles.

1. Understand that these limits — whether for anatomic measurements, laboratory test values, or physiologic measurements — refer to the upper and lower percentile points on a frequency distribution of such measurements made on "clinically normal" persons.

2. Understand that the patient population upon which such normal limits were derived should have similar characteristics to patients from whom data will be compared.

IV. Demonstrate an understanding of sampling variation and confidence intervals.

A. Explain sampling variation for percentages and measurements.

B. Given a standard error (S.E.) calculated from a set of measurements, state its meaning and contrast this with a calculated standard deviation.

C. Given a confidence interval for a mean or a percentage, state the appropriate interpretation.

V. Demonstrate an understanding of tests for statistical significance.

A. For the following three situations, given a statement concerning the statistical significance of the results of hypothesis testing, provide the appropriate interpretation, including a probability statement:

1. A chi-square test used to compare two percentages.
2. A paired (one-sample) **t** test.
3. An unpaired (two-sample) **t** test.

B. Precisely state what a **p** value (**p** < or **p** >) means in such tests.

C. State the two types of errors found in significance testing.

D. Explain the meaning of test power and its relation to sample size.

E. Contrast practical significance and statistical significance.

VI. Demonstrate an understanding of simple linear regression and correlation.
   A. Given a set of paired measurements (two variables measured on **n** items or subjects), graphically represent the relationship between the variables by constructing a scatter diagram (scattergram).
   B. Given a regression equation calculated from a specified data set (from an experiment or nonexperimental study), interpret the meaning of the equation.
      1. Distinguish between linear and nonlinear association.
      2. State the meaning of the true intercept **A** and its calculated estimate **a**.
      3. State the meaning of the true slope **B** and its calculated estimate **b**.
   C. Given the results of a statistical significance test of the hypothesis that the slope **B** = 0, interpret these in terms of a probability statement.
   D. Explain the meaning of a calculated linear correlation coefficient, **r**, in terms of
      1. Knowledge that the utility of a correlation coefficient is limited to a nonexperimental study.
      2. Distinction between correlation and causation.
      3. Distinction between positive and negative correlations.
      4. Distinction among zero, moderate, and perfect correlations.
   E. Given linear correlation coefficients (**r**'s) calculated from specified data sets, interpret the meaning of the coefficients with respect to the relevant populations.
   F. Given the results of a statistical significance test of the hypothesis that the true correlation between two variables is zero, interpret these in terms of a probability statement.

G. Given coefficients of determination (R^2) calculated from specific data sets, precisely interpret the meaning of them with respect to the relevant populations.

VII. Understand the rationale for clinical trials.
   A. Specify the major difference between inferences that may be drawn in experimental studies and those in nonexperimental studies.
   B. Define an experiment and list the fundamentals of experimentation.
   C. Understand the basic concepts of clinical trials.
      1. Define clinical trial.
      2. Define single-blind clinical trial.
      3. Define double-blind clinical trial.

VIII. Understand the basic concepts of epidemiology.
   A. Define epidemiology.
   B. Describe criteria by which epidemiologic evidence may be judged in attempting to assess causality.
   C. Define a case-control study.
   D. Define a cohort study.
   E. Define a risk factor.
   F. Differentiate between absolute risk and relative risk.
   G. Describe confounding.
   H. Calculate and interpret a relative risk from cohort study data summarized in a 2×2 table.
   I. Calculate and interpret an odds ratio from case-control study data summarized in a 2×2 table.
   J. Know when an odds ratio may well approximate a relative risk estimate.
   K. Understand how basic statistical methods are used to assess observed associations between suspected risk factors and disease.

Instructions

The following two general rules for using this book will help you to achieve the best results:

1. Conceal the answers for each section until you have written down your answer.
2. Use short study periods; 20- to 30-minute sessions should be optimum.

Many of the responses are trivial or repetitive in terms of "recall" requirements. However, most students find it advantageous to write out **all** the answers in the spaces provided.

Descriptive
Statistics

1

FUNDAMENTAL CONCEPTS

1.1 If a set of numerical observations, when plotted on a number scale, may lie only on certain isolated points, and not on the points in between, then the set is called a set of **discrete data**. If the set of observations may theoretically lie anywhere within a specified interval on the number scale, the set is called a set of **continuous data**. The process of **counting** produces discrete data; the process of **measurement** yields continuous data. A set of observations of the body temperatures of patients would therefore be a set of _____ data.

continuous

1.2 Graphically, the above concept may be displayed as shown:

Continuous data:

In this case, **any** value between points **a** and **b** in a "relevant range" is theoretically possible.

Discrete data:

Here, only _____ points are possible.

isolated, single

1.3 A letter, such as **X** or **Y**, that is used to represent a member of a set of discrete data is usually called a **discrete variable**. Similarly, such a letter used to represent a member of a set of continuous data is called a **continuous variable**. Thus, if **Y** represents "number of children in a family," it is a _____ variable.

discrete

1.4 In a person whose systolic blood pressure has been known to vary between 130 and 140 mmHg,

130 mmHg   140 mmHg

any value is possible between 130 and 140 mmHg, even 137.4329. Because of the limitations of our measuring devices, however, we may be forced to round off observations to three or four significant digits. Despite this practical limitation, a measurement Y, such as blood pressure, is a _____ variable. Continuous data are sometimes called "measurement data."

1.5 Numerical observations may comprise either discrete or continuous data. The procedure of measuring students' heights would yield _____ data; the process of counting the number of persons who have contracted a specific disease would yield _____ data.

1.6 A **percentage** is computed on the basis of a ratio of a count ($\times$ 100) to a total number. Thus, the percentage of patients possessing a certain attribute is a member of a set of _____ data.

1.7 Discrete or **categorical** variables may be only **nominal**, such as male or female, or white, black, or Asian. However, when categorical variables have an order to them, they are called **ordinal** variables, such as mild, severe, and fatal (when referring to the severity of a disease). Therefore, sex and ethnicity are _____ variables, and the intensity of pain evaluated as $+$, $++$, or $+++$ is a(n) _____ variable.

1.8 Discrete or categorical variables that have only two different values, such as positive-negative or alive-dead, are called **dichotomous**. Exposure status (exposed-unexposed) and outcome (diseased-well) are examples of _____ variables. Exposure status, in terms of a person's current cigarette-smoking status, could be graded as none, low, or high. This is another example of a(n) _____ variable.

It is important to distinguish between discrete data (such as counts and percentages) and continuous data (measurements) because statistical methods for description and analysis are usually different, as will be illustrated throughout this book.

continuous

continuous or
measurement
discrete

discrete

nominal

ordinal

dichotomous

ordinal
(This is also an
example of a cate-
gorical variable.)

POPULATIONS AND RATES

1.9 From a statistical standpoint, a **population** is the totality or set of elements that has one or more characteristics in common. A population is **specified** when the common characteristics that define it have been specified. According to such a definition, we may refer not only to the inhabitants of a region but also to several other types of populations as well. The kinds of populations may be divided into populations of **tangible things** and populations of **conceptual values**.

An example of the first kind of population is illustrated by the total serum cholesterol values for all healthy 21-year-old men in the state of Iowa. Another is the dry weights of right kidneys in adult male chimpanzees in a specified colony.

An example of the conceptual or hypothetical kind of population is all possible systolic blood-pressure readings made in one person by the same observer under controlled conditions. Another is all possible average (mean) values of serum cholesterol values that one might obtain from 10 experimentally treated rabbits selected from a group of rabbits.

Thus, a _____ is the set of elements that has one or more characteristics in common.

population

1.10 **Counts**, **ratios**, **proportions**, and **percents** are used to summarize characteristics or attributes in a population or a subset of a population. Simple **counts** are useful for evaluating the absolute magnitude of a characteristic in the population. Thus, the number of patients who possess an attribute (such as being female or being a current smoker) is just a _____. However, this number only assesses the absolute frequency of an attribute. **Ratios** are used to assess the relative importance of a characteristic in comparison to those without the attribute or characteristic; the numerator is not included with the denominator. The male-female ratio is an example.

count

Number of males:	60	Male-female ratio:
Number of females:	40	$60:40 = 1.5$ or $1.0:1.5$

Total	100

Proportions are statistics in which the numerator must be contained in the denominator. Proportions assess the relative importance of an attribute in the population.

Percents (or percentages) are proportions multiplied by 100. Therefore, in the previous example, there are $(^{60}/_{100}) \times 100 = 60$ _____ males in the population. However, the male-female _____ is 1.5 or 1.0:1.5.

percent (%)
ratio

1.11 In biostatistics the term **rate** is usually used to specify the frequency of a phenomenon in a defined population. For example, the rate or percentage of a population possessing an attribute **a** is determined by the ratio.

$$\frac{\text{Number possessing } \mathbf{a}}{\text{Total number in population, } \mathbf{N}} \times 100$$

Two important disease rates that are useful in public health studies are **prevalence** and **incidence**.

$$\textbf{Prevalence rate} = \frac{\text{number of cases at a given point in time}}{\text{number of persons in the population at that time}}$$

Usually, this ratio is multiplied by a constant value, such as 100 (to yield percentage) or 100,000.

$$\textbf{Incidence rate} = \frac{\text{number of new cases occurring during a given time period}}{\text{total (or average) number of persons in population during same period}}$$

The time period mentioned in this equation is usually 1 year.

A group of 10 persons was observed for 1 year. Six became ill with disease A at some time during the year. Their length of illness is represented in the table that follows by a line (_____).

	Month											
Person no.	J	F	M	A	M	J	J	A	S	O	N	D
1		—										
2									—			
3												
4				—								
5												
6					—							
7				—								
8												
9			—									
10												

The incidence rate for this population for the year was _____%. The prevalence rate on February 1 was _____%, and the prevalence rate on March 15 was _____%.

60
30
50

When computing incidence or prevalence rates, the **population** to which the rates refer must be carefully specified. Ordinarily, separate rates are calculated by sex and by age groups. Sometimes these rates are determined for other subpopulations as well, depending on the purposes of the study.

1.12 Suppose a particular diagnostic test yields a result such that a patient is classified as positive (disease present) or negative (disease absent). Few such tests will always correctly classify diseased patients and well patients; that is, for one reason or another, there will inevitably be false positives and false negatives. For a particular population in which each member is tested once for a particular disease, we may classify persons as follows:

Disease state

Test result	Number of persons with	
	Disease	*No disease*
Positive	True positives	False positives
Negative	False negatives	True negatives

Two rates that are often useful in the evaluation of such diagnostic tests are the **sensitivity** rate and the **specificity** rate.

$$\textbf{Sensitivity} = \frac{\text{number of true positives}}{\text{number of true positives} + \text{number of false negatives}} \times 100$$

which is the percentage of diseased persons detected (classified positive) by the test.

$$\textbf{Specificity} = \frac{\text{number of true negatives}}{\text{number of true negatives} + \text{number of false positives}} \times 100$$

which is the percentage of well persons classified as well (negative) by the test.

The following example illustrates the computation of these rates:

Disease state

Test result	Number of persons with	
	Disease	*No disease*
Positive	200	100
Negative	40	900
Total	240	1000

$$\text{Sensitivity} = \frac{200}{200 + 40} = \frac{200}{240} = 83\%$$

$$\text{Specificity} = \frac{900}{900 + 100} = \frac{900}{1000} = 90\%$$

Another way of interpreting these terms is that

Sensitivity = 1 − false negative rate (proportion)

and

Specificity = 1 − false positive rate (proportion)

where the false negative rate is the proportion of true diseased persons diagnosed as negative (40/240 in the example) and the false positive rate is the proportion of true nondiseased persons with a positive test result (100/1000 in the example).

The following data were obtained in a diagnostic test evaluation study:

Disease state

Test result	*Number of persons with*	
	Disease	*No disease*
Positive	80	10
Negative	20	190

The test's sensitivity is estimated to be _____%, and the specificity is estimated to be _____%.

80
95

Since no diagnostic test is perfect, the sensitivity and specificity of new tests must be carefully assessed. It is desirable to have both high sensitivity and specificity. However, it is usually impossible to maximize both rates, and choices must be made in establishing the test criteria for positive and negative results. Such choices will depend on the relative costs (in a general sense) of obtaining a false positive result and the costs of a false negative result.

DESCRIPTIVE MEASURES

1.13 Before a set of observations can be used as a basis for inference, they usually must be concisely described or summarized. One way of summarizing a group of measurements or counts is by the use of a **frequency distribution**. The range of all values is divided into ordered classes, and the number of observations that fall into each class is determined. When the total number of observations is small, the numerical values themselves may be used to define the classes.

Complete the frequency distribution in the table of the following 10 measurements: 3, 6, 5, 6, 5, 2, 5, 4, 5, 4:

Measurement	Number of measurements (frequency)
2	1
3	1
4	2
5	4
6	
Total	10

1.14 A laboratory determined the sodium concentration (mEq/L) on aliquots of a single specimen of serum on 30 different days. The values obtained were

143	142	140	144	142	141
142	143	144	144	145	145
141	142	143	143	141	141
142	144	144	142	143	143
143	146	143	144	144	145

Construct the appropriate frequency distribution in the space provided in this table:

Determination	Number of determinations		Det.	No. of det.
			140	1
			141	4
			142	6
			143	8
			144	7
			145	3
			146	1
			Total	30

1.15 A **relative frequency distribution** may be obtained by dividing the absolute frequencies by the total number of observations. The relative frequency distribution and the percentage frequency distribution for the 30 sodium determinations are

Determination	Relative frequency	Percentage frequency
140	.033	3.3
141	.133	13.3
142	.200	20.0
143	.267	26.7
144	.233	23.3
145	.100	10.0
146	.033	3.3
Total	.999	99.9

In the space below, construct the relative frequency distribution and percentage frequency distribution for the 10 measurements given in section 1.13.

Measurement	Relative frequency	Percentage frequency	Meas.	Rel. freq.	% freq.
			2	.10	10
			3	.10	10
			4	.20	20
			5	.40	40
			6	.20	20
			Total	1.00	100

1.16 A **cumulative frequency distribution** may be constructed from the sodium determinations as follows:

Determination	Number of determinations	Cumulative relative frequency
139 or less	0	.000
140 or less	1	.033
141 or less	5	.166
142 or less	11	.366
143 or less	19	.633
144 or less	26	.866
145 or less	29	.966
146 or less	30	.999

Note that the number of determinations is cumulatively added $(1, 1 + 4, 1 + 4 + 6,$ and so on) until all are accounted for. The corresponding cumulative relative frequencies (or just cumulative frequencies) may be obtained by dividing these cumulative totals by the total number of determinations, **n**, which, in this case, is 30. Of course, rounding-off error is possible in the last decimal place.

Construct the cumulative frequency distribution for the 10 measurements given in section 1.13.

Measurement	Number of measurements	Cumulative relative frequency

Meas.	No. of meas.	Cum. rel. freq.
1*	0	0
2*	1	.10
3*	2	.20
4*	4	.40
5*	8	.80
6*	10	1.00

*Or less.

1.17 A graphic representation called a **cumulative frequency polygon** may be constructed from the table in 1.16 as shown in the following graph:

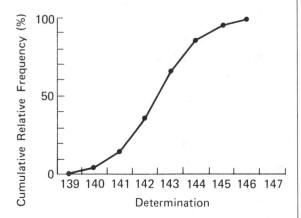

From this graph, it can be seen that approximately _____% of the measurements are less than or equal to 144.

85

1.18 A **histogram** is a frequency distribution shown in bar graph form. Each bar is made proportional to the frequency of the observations in each class (grouping) of observations. The **relative frequency histogram** for the 10 measurements in our example (sections 1.13 and 1.15) is shown.

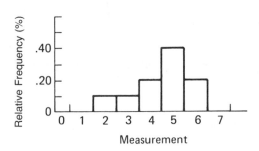

Construct the relative frequency histogram for the 30 sodium determinations:

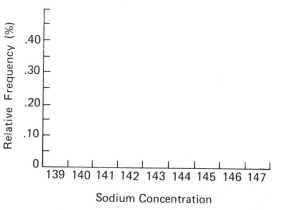

Sodium Concentration

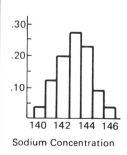

Sodium Concentration

NOTE: Histograms constructed with class intervals of **unequal width** require special attention. In all cases, it is the **area** of the bar that represents the frequency. See Reference 1 (p. 21) for details.

1.19 A frequency distribution or a histogram gives a good picture of the pattern or distribution of the observations. However, further summarization is usually necessary, particularly before inferences or generalizations are drawn from the data. Methods or measures for describing the "middle" (central tendency) of a set of observations and their "spread" (variation) are used for this purpose.

Before discussing other specific descriptive measures, it will be useful to introduce the concept of an **array**. An array of a set of numbers is simply those numbers in (algebraically) ordered sequence from smallest to largest; for example, the array for the set 2, 18, 3, −4, 1, 2, 6 is

−4, 1, 2, 2, 3, 6, 18

Thus, the following sequence of numbers—2, 4, 4, 5, 7, 9—is also a(n) _____ .   array

1.20 An array of a set of **n** measurements may be written algebraically as $X_1, X_2, \ldots, X_n$. Their sum may be written as

$$\sum_{i=1}^{n} X_i = \Sigma X = X_1 + X_2 + \ldots + X_n$$

where Σ (capital sigma) is the summation sign. In the set of measurements in section 1.19, $\mathbf{n} = 7$, $\mathbf{X_1} = -4$, $\mathbf{X_2} = 1$, $\mathbf{X_3} = 2$, $\mathbf{X_1} = 2$, and so on. In the same set, $\mathbf{X_7} = $ _____ and

18

$$\sum_{i=1}^{n} \mathbf{X_i} = \Sigma\mathbf{X} = \text{_____}$$

28

1.21 Three measures used to describe the **central tendency** of a set of observations are the mean, median, and mode. The arithmetic **mean** of a set of **n** numbers is the simple **average**, algebraically expressed as

$$\overline{\mathbf{X}} = \frac{1}{\mathbf{n}}\Sigma\mathbf{X_i}$$

The mean, $\overline{\mathbf{X}}$, of the seven numbers in 1.19 is $28/7 = 4.0$. The mean of the following array—5, 6, 8, 10, 11—is _____ .

8

1.22 The **median** of a set of **n** numbers is the "**middle-ranking**" **number**. The set must first be ordered. If **n** is an **odd** number, then the [(**n** + 1)/2] nd number in the array (ordered set) is the median. Consider the array 1, 2, 2, 3, 4, 6, 9. Since **n** = 7, the median is the $(7 + 1)/2 = $ 4th number in the array, or 3. Thus, the median of the array 6, 12, 76 is _____ , and the median of the set 11, 13, 7, 8, 4 is _____ (do not forget to order the set first).

12
8

1.23 If **n** is an **even** integer such as in the array 2, 3, 3, 4, 7, 8, 9, 14, we find (**n** + 1)/2 = (8 + 1)/2 = 4.5. Any value between the 4th and 5th numbers may be called the median, that is, any value between 4 and 7. It is customary, however, to choose the midpoint of this interval. Hence, the median is 5.5.

Obviously, the median is less sensitive to extreme observations than the mean. This is especially true for small sets of numbers.

The median in the following array of measure-

ments—1, 2, 3, 3, 3, 6, 9, 10, 11, 19, 21, 22, 47—is
_____ .

9

1.24 The median in the following array of measurements—7, 8, 10, 15, 19, 25—is _____ .

12.5

1.25 The **mode** is the most frequently occurring number in a set of observations. The mode in the set 1, 2, 2, 4, 4, 4, 6, 17 is 4. Because a mode may not be unique, it is not a particularly good measure of central tendency. The mode of the first and second sets in section 1.23 is _____ .

3

1.26 Different groups of observations may have the same mean, median, or mode yet differ considerably with respect to the **spread among the individual observations**. Some descriptive measures of this **variation** are therefore necessary. The **range**, **standard deviation**, and **variance** are such measures.

The **range** of a set of numbers is simply the difference between the largest and smallest values. The range of the set of numbers in section 1.24 is _____ . When reporting the range, it is usually desirable to include the values of the largest and smallest numbers. Thus, in this case, we would report the range as _____ to _____ .

18

7 to 25

1.27 The fact that the range utilizes information from only two of the available observations makes it a somewhat inefficient measure of variation, particularly with large numbers of observations. Obviously, the range will be related to this number. A measure that does not have this shortcoming is the **variance**, s^2, and the square root of the variance, the **standard deviation**, **s**.

Consider a set of observations $X_1, X_2, \ldots, X_n$ with a mean, $\overline{X}$. The variance may be calculated by the following formula:

$$s^2 = \frac{1}{n - 1} \Sigma(X_i - \overline{X})^2$$

Thus, for the following set of measurements 5, 15, 15, 20, and 45, $n = 5$, $\Sigma X_i = 100$, $\overline{X} = 20$, and

$X_i - \overline{X}$	$(X_i - \overline{X})^2$
$5 - 20 = -15$	225
$15 - 20 = -5$	25
$15 - 20 = -5$	25
$20 - 20 = -0$	0
$45 - 20 = +25$	625
Sum $\overline{0}$	$\overline{900}$

The variance $s^2 = $ _____. The standard deviation $s = $ _____.

$$\frac{900}{4} = 225$$
$$\sqrt{s^2} = \sqrt{225} = 15$$

1.28 An algebraically equivalent, and usually simpler, formula for calculating the variance is

$$s^2 = \frac{1}{n-1}\left\{ \Sigma X_i^2 - \frac{(\Sigma X_i)^2}{n} \right\}$$

Using the figures from the preceding example, we have:

X_i	X_i^2
5	25
15	225
15	225
20	400
45	2025
100	2900

Therefore,

$$s^2 = \frac{1}{4}\left\{ 2900 - \frac{(100)^2}{5} \right\} = \text{_____}$$

225

NOTE: A large variance (or its corresponding standard deviation) generally indicates that the observations have more "spread" than a group in which the variance is smaller when the units of measurement are identical. However, such conclusions depend to a degree on the shape of the frequency distributions. A specific interpretation can be given to **s** if the data appear to be "normally" distributed; that is, if a mathematical model called the "Gaussian" or "normal" distribution seems to describe the data. This distribution is discussed in Chapter 3.

1.29 In contrast to the calculated deviation, **s**, the **coefficient of variation** is a measure of the relative, rather than absolute, variation. The coefficient of variation, **C.V.**, is simply the ratio, $s/\overline{X}$, (often multiplied by 100). Because both **s** and $\overline{X}$ are given in the same units of measurement, the C.V. is a dimensionless quantity. One use is to compare the relative variation between different kinds of measurements. The coefficient of variation for the set of measurements given in section 1.27 is _____.

.75 or 75
$$\left(\frac{15}{20}\right)$$

1.30 Percentiles provide a way of describing the variation of a frequency distribution regardless of the shape of the distribution. **Percentiles** are numbers that divide a distribution or the area of a histogram into 100 parts of equal area. The 10th percentile, for example, exceeds 10% and is exceeded by 90% of the observations. The 75th percentile exceeds _____% of the data.

75

1.31 The **median** is the _____ percentile.

50th

1.32 In the frequency distribution for the 20 values shown in the figure, the 20th percentile value is _____.

3.0; it exceeds exactly four-twentieths of the observations.

The 95th percentile is_____.

9.5, or halfway

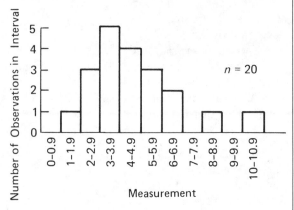

between the 19th and 20th numbers in the array. This is determined by simple linear interpolation: any point between the two highest-ranking observations will separate the upper 5% from the other 95% when **n** = 20 (1/20 = 5%).

1.33 For the following cumulative frequency polygon, the 10th percentile is approximately _____, the 50th percentile is approximately _____, and the 90th percentile is approximately _____.

3.0
8.0
18.0

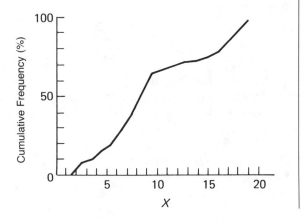

Probability

2

PROBABILITY OF AN EVENT

2.1 A **random process** or **random experiment** is a repetitive process or operation that, in a single trial, may result in any one of a number of possible outcomes such that the particular outcome is determined by chance and is impossible to predict. We cannot know in advance of a toss whether a penny will fall heads or tails, nor can we predict in advance what number will occur when a pair of dice is rolled. The fact that the outcome cannot be predicted is due to the element of **chance** or **randomness**.

If a random process can result in n equally likely and mutually exclusive outcomes and if a of these outcomes has an attribute A, then the probability of A is the ratio a/n:

$$P(A) = \frac{a}{n}$$

Two outcomes are **mutually exclusive** if, when one occurs, the other cannot. A **mutually exclusive and exhaustive** set is a set of outcomes such that one of the set must occur, but only one can occur in a single trial.

When a die is rolled, either an odd or even number must appear. These two possible results are said to be _____ _____.

mutually exclusive

2.2 When it is stated that the probability that a 25-year-old white United States citizen will not survive 1 year is .008, the interpretation is that the relative frequency (incidence) of death in this population in 1 year has been 8/1000. The statement that the probability of surviving a particular surgical operation is .90 means that the _____ _____ of occurrence of death with many such operations is 10% in a specified time period.

relative frequency

2.3 A patient may be classified as having coronary heart disease, diabetes, or hypertension. These events _____ (are, are not) mutually exclusive.

are not

2.4 A patient may be classified as having acute, chronic, or no appendicitis. These three events _____ (are, are not) mutually exclusive.

are

2.5 The above classical definition of probability (section 2.1) is useful when all possible outcomes (events) of a procedure are equally likely. For example, the probability of obtaining a head when tossing a fair coin is one-half, since the number of possible outcomes = 2, and the number of outcomes having attribute **a** (heads) = 1. Similarly, the probability of drawing an ace from a deck of cards is _____ .

$$\frac{4}{52} = \frac{1}{13}$$

2.6 The classical definition of probability does not apply, however, in situations where all events are not equally likely to occur.

A second, broader definition of probability is based on the fact that if a random experiment, some of whose outcomes have an attribute **A**, is repeated an indefinite number of times, the relative frequency of the occurrence of **A** tends to converge to a constant value. Suppose we conduct a sequence of experiments such that the **i**th experiment consists of **n**$_i$ trials of the event. In the **i**th experiment, **A** occurs **a**$_i$ times; therefore, the relative frequency of **A** in this experiment is the ratio **a**$_i$/**n**$_i$. If we combine the results of the experiments as they are performed, the relative frequency **f** at the end of the **k**th experiment is

$$\mathbf{f} = \frac{\mathbf{a}_1 + \mathbf{a}_2 + \ldots + \mathbf{a}_k}{\mathbf{n}_1 + \mathbf{n}_2 + \ldots + \mathbf{n}_k} = \frac{\Sigma \mathbf{a}_i}{\Sigma \mathbf{n}_i}$$

Although in any such sequence the individual relative frequencies may be very different, we find that, as more and more experiments are performed, the combined frequency ratio **f** soon becomes relatively stable and, in the long run, tends to converge to some value that we call the **probability of** A.

We may illustrate this property by tossing a coin. If the coin is fair, the classical definition tells us that the probability of a head is equal to one-half; that is, if we toss the coin a number of times, we would expect to obtain a head on about one-half the tosses. In any given number of tosses, the relative frequency of heads may differ from one-half, but, in the long run, if we were to continue tossing the coin, we would expect the relative frequency to approach that value. The results of an experiment of this sort are plotted on a semi-logarithmic scale in the figure below. Here we can plainly see the convergence of the frequency ratio to the value _____.

$^{1}/_{2}$

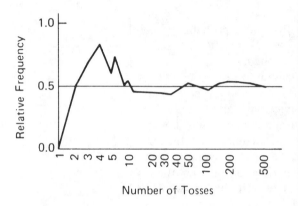

Number of Tosses

2.7 Given a random process for which the relative frequency of occurrence of **A** seems to display this statistical stability (i.e., seems to converge to a constant value), we will postulate the existence of a number, **P**, which is the probability for the outcome **A**. For any **finite** sequence of trials of the experiment, the observed **relative frequency a/n** of **A** is an **estimate** of the probability of **A**. The probability of an event **A** may be said to be the _____ _____ of the occurrence of **A** in a series of trials.

relative frequency

2.8 If the number of trials is large, we can reasonably expect that the observed relative frequency will be close to the **theoretical relative frequency** or the true probability of **A**. It is evident that if we are to define probability as the theoretical relative frequency, we can logically speak of probabilities only when the processes under consideration are _____ (repetitive, non-repetitive).

repetitive
(at least
conceptually)

SOME BASIC RULES CONCERNING PROBABILITIES

2.9 Using the above relative frequency definition of probability, it is obvious that any probability must be a number (a proportion) between zero and one, inclusively. The probability of an event, **A**, is written **P(A)** and must satisfy the following condition:

$$0 \leq P(A) \leq 1$$

Thus, we can say that a certain event has probability one and that an impossible event has probability _____ . This defines the complete range.

zero

2.10 If **A** and **B** are two events that may result from a single repetition of a random process, the probability that **A** or **B**, or both, will occur is the probability of **A** plus the probability of **B** minus the probability that both **A** and **B** occur. Symbolically:

$$P(A \text{ or } B) = P(A + B)$$
$$= P(A) + P(B) - P(AB)$$

If **A** and **B** are mutually exclusive outcomes, then

$$P(A + B) = P(A) + P(B)$$

This is because **P(AB)** must equal _____ when **A** and **B** are mutually exclusive.

zero
Here **AB** is an impossible event.

2.11 Using the above formula, the probability of obtaining an ace or a king in a single, randomly drawn card from a deck is _____ .

$\dfrac{4}{52} + \dfrac{4}{52} = \dfrac{8}{52} = .154$
Ace and king are mutually exclusive outcomes.

2.12 The probability of drawing an ace or a heart in a single draw is _____ .

$$\frac{4}{52} + \frac{13}{52} - \frac{1}{52}$$
$$= \frac{16}{52} = .308$$

Ace and heart are not mutually exclusive outcomes.

2.13 Suppose that we now have the results of two repetitions of a random process. Now we say that the probability that the joint occurrence of two events, **A** and **B**, is the product of the probability of **A** and the **conditional probability** of **B**, given that **A** has occurred. Equivalently, we can state that this is the probability of **B** multiplied by the conditional probability of **A**, given that **B** has occurred. Symbolically:

$$P(AB) = P(A) P(B|A) = P(B) P(A|B)$$

If **A** and **B** are **independent** events (the occurrence or nonoccurrence of one does not affect the probability of occurrence of the other), then the formula simplifies to

$$P(AB) = P(A) P(B)$$

since $P(B|A) =$ _____ in this important case.

$P(B)$
The occurrence or nonoccurrence of **A** on the first "trial" is immaterial to results on trial two.

2.14 If two consecutive cards are drawn from a deck without replacement, the probability $P(AB)$ of obtaining an ace (**A**) on the first draw and a king (**B**) on the second is

$$P(AB) = P(A) P(B|A)$$

$$P(AB) = \left(\frac{4}{52}\right)\left(-\right) = .0060$$

$$\frac{4}{51}$$

The probability of obtaining a king on the second draw given that an ace was obtained on the first draw.

2.15 Suppose two cards are consecutively drawn from a deck but after the first draw, the card is replaced and the deck shuffled before the second draw. Now what is the probability of obtaining an ace on the first draw and a king on the second?

$$P(AB) = \underline{\hspace{2cm}}$$

$$\left(\frac{4}{52}\right)\left(\frac{4}{52}\right) = .0059$$

These two events are independent since the outcome of draw two is in no way influenced by the result of draw one.

2.16 If two coins are tossed, the probability that both are heads is _____ .

$$\left(\frac{1}{2}\right)\left(\frac{1}{2}\right) = \frac{1}{4}$$

These are also independent events.

If two dice are rolled, the probability of obtaining a one on each is _____ .

What is the probability that the first two children in any given family will be boys? _____

$$\left(\frac{1}{6}\right)\left(\frac{1}{6}\right) = \frac{1}{36}$$

$$\left(\frac{1}{2}\right)\left(\frac{1}{2}\right) = \frac{1}{4}$$

The assumption of independence and a sex ratio at birth of $1:1$ is very close to the truth in nature.

2.17 The probability of contracting diabetes mellitus during one's lifetime is approximately .04, and the probability of developing some form of cardiovascular disease is approximately .50. If these two elements are independent, the probability of one person contracting both during his or her lifetime is approximately _____ . (Incidentally, these probabilities are **not** independent.)

.02
(.04) (.50)

2.18 Conditional probability is an extremely important concept involved in statistical decision-making in a wide variety of contexts. For one example, consider again the 1000 persons displayed in the 2×2 table of section 1.12. From this table we can see that the probability of having a positive test result can be said to be a conditional probability, depending on whether or not the disease is truly present in the person being tested for this disease.

If **D** signifies disease, $\overline{D}$ means no disease, and $+$ means a positive test result, we could write the conditional probabilities

$$P(+|D) = .83,$$
$$\text{and } P(+|\overline{D}) = \text{_____}$$

.10
This is the false positive rate, that is, the probability of a positive test result given that the person selected is disease free.

2.19 From this same table of 1000 subjects and using the conditional probability concept, we can restate (as above) that the probability of obtaining a $+$, conditional on having the disease, is the sensitivity of the test. Analogously, the probability of obtaining a $-$, conditional on not having the disease, is called the _____ of the test. In this particular table,

specificity

$$P(-|\overline{D}) = \text{_____} .$$

.90

2.20 Another useful set of rates, which can be expressed as conditional probabilities, is based on the data available in this same kind of 2×2 table. These include the probability that a subject who tests positive is actually diseased and the probability that a person with a negative test re-

Probability

sult is disease free. Using the abbreviations **TP** = the number of true positives, **FP** = false positives, **TN** = true negatives, and **FN** = false negatives, we can write:

$$\text{Positive predictive value} = \frac{TP}{TP + FP}$$

and

$$\text{Negative predictive value} = \frac{TN}{TN + FN}$$

n the table of section 1.12, we can see that the positive predictive value equals

$$\frac{200}{200 + 100} = 0.67$$

This is the probability of having the disease given a positive test; or $P(D/+) = 0.67$.
The negative predictive value = _____.

$$\frac{900}{900 + 40} = 0.96$$

This is the probability of being disease-free given a negative test; or $P(\overline{D}/-)$ in this data set.

Population Distributions and Samples

3

POPULATIONS AND SAMPLES

3.1 For a measurement made on a single patient (or subject) to have the most meaning, it is desirable to compare it with the distribution of all such measurements made on the complete population of all well and diseased persons in the same categories (sex, age group, race, geographic area, and so on) as the patient. It is obviously impossible to obtain such data on such complete populations; therefore, investigators must be content with data from what is hoped to be a representative subset called a **sample**.

In this sense, **bias** may be defined as something that makes a sample different from what it purports to be (see References 1 and 15). For example, data from a group of male hypertensive outpatients at a single clinic were summarized with the implication that they represented all male hypertensives. Such a group would be a(n) _____ sample.

biased (For a variety of reasons, clinic populations usually differ from each other and from the general population. Location and other factors can affect referral patterns.)

3.2 When sampling from an infinite (or a very large finite) population, a **random sample** is defined as one selected in such a way that **every element in the population has an equal and independent chance of being selected**. Thus, a single card drawn from a well-shuffled deck could be considered a(n) _____ sample of size _____ .

random
one

3.3 A questionnaire was administered only to morning patients at a clinic. This group could not be considered a random sample of clinic patients, because each patient did not have a(n) _____ chance of being included.

equal

3.4 Volunteers from a particular patient population participated in an experiment to evaluate two therapies clinically. The results of such a study cannot be safely generalized to the entire patient population, because the sample was not _____. Patients who volunteer may differ from those who do not volunteer with respect to important factors. Such a sample is _____.

random

biased

NOTE: In nearly all experimental studies involving humans, this is a constraint that must be borne in mind when interpreting the resulting data.

3.5 One other type of sampling is called **systematic**. Such a sample is one selected according to some system, such as choosing every 10th patient for a more extensive examination. It is difficult to ascertain that some bias with the use of the same system does not exist. In some cases, a systematic sample may be more representative than a random sample, but, unfortunately, systematic sampling has proved to be difficult to treat mathematically. The two major advantages of random sampling from a population are (1) the chances for bias are minimized and (2) probability statements may be employed in the evaluation of results. A _____ (random, systematic) sample is less likely to be biased.

random

POPULATION DISTRIBUTIONS

3.6 If the size of a random sample from a continuous-variable population is indefinitely increased and the measuring instrument is such that smaller and smaller intervals may be used, the histogram will "approach" a smooth curve called the **population distribution** or **distribution**. An example is shown below where the variable **X** that is being measured is the fasting serum triglyceride level. One value each was obtained from a sample of men.

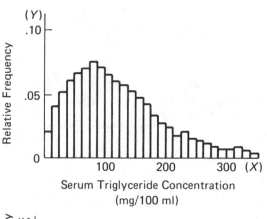

Serum Triglyceride Concentration
(mg/100 ml)

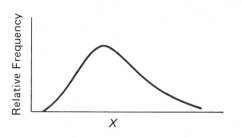

Serum Triglyceride Concentration
(mg/100 ml)

The following smooth curve for a (continuous) measurement variable **X** is called a(n) _____ .

population distribution

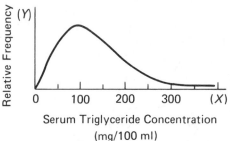

THE NORMAL (GAUSSIAN) DISTRIBUTION

3.7 The above "limiting form" of an empirically determined histogram may also be called a **theoretical frequency distribution**. Several mathematical models are often used to approximate such distributions for continuous data. Using **Y** to represent relative frequency, such a model expresses **Y** as a function of the measurement variable X; that is, $Y = f(X)$. The most commonly used model for population distributions is the **Gaussian distribution** (after the mathematician J. C. F. Gauss) or **normal distribution**. This model is used to describe measurements that have an approximately **symmetric** bell-shaped histogram, as shown below:

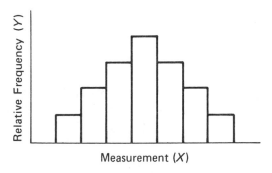

Another name for a normal distribution is a(n) _____ distribution.

Gaussian

3.8 The general mathematical form of the normal distribution is

$$Y = \frac{1}{\sigma\sqrt{2\pi}} \, e^{-\frac{1}{2\sigma^2}(X-\mu)^2}$$

where X is the measurement variable and Y is the frequency value corresponding to X. This distribution depends on two **parameters** (i.e., constants that must be specified in order to designate a particular member of a family of distributions of the same general form); these are the **mean**, μ, and the **variance**, σ^2. The square root of the variance is called the **standard deviation**, σ. Symbols representing the two parameters of a normal distribution are _____ and _____.

μ, σ^2

3.9 Both parameters μ and σ^2 must be given numerical values to define a particular normal distribution from the general family. If μ is fixed but σ^2 is not, we have an infinite number of distributions that have the same mean but different variances, examples of which are shown in

Figure A, below. If σ^2 is fixed, but not μ, we have an infinite number of distributions with the same shape but with different locations along the X axis, as illustrated below in Figure B.

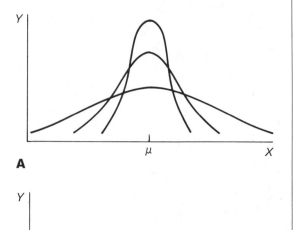

A

B

The following two normal distributions have the same _____ but different _____.

variance (σ^2),
means (μs)

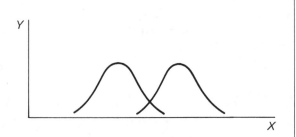

3.10 The following two normal distributions have the same _____ but different _____.

mean (μ),
variances (σ^2s)

Population Distributions and Samples

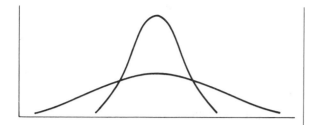

3.11 The mean μ and the variance σ^2 are population parameters that are estimated from a sample from the population. The sample mean $\overline{X}$ and the sample variance s^2 provide estimates of μ and σ^2. Also, s is the estimate for σ. The formulas for obtaining $\overline{X}$ and s^2 were presented in Chapter 1. As was the case for sample variances, population variances are a measure of "spread" or variation; a larger σ^2 indicates greater spread. In the following figure, curve _____ (A, B) has the larger σ^2.

B

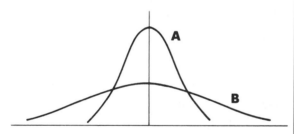

3.12 There is a very convenient way of using the standard deviation σ to specify the spread of measurements in a normal distribution. It can be shown that for any normal distribution, the following relationships hold:

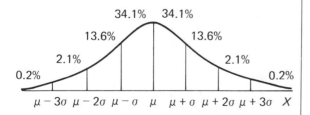

Thus, we can say that the interval $\mu \pm \sigma$ contains 68% of the population values, $\mu \pm 2\sigma$ contains about _____% of the population values, and μ

95

$\pm 3\sigma$ contains about _____ % of the population values.	99.6

NOTE: In fact, $\mu \pm 1.96\ \sigma$ contains exactly 95% of the values. We have rounded to the whole number 2.

3.13 In one particular normal distribution, suppose that $\mu = 0$ and $\sigma = 1$. In this case, _____ % of all population values lie between -2 and $+2$.	95

3.14 In another normal distribution where $\mu = 100$ and $\sigma = 10$, about 68% of all population values lie between _____ and _____ .	90, 110

3.15 Approximately the same relationship as above will hold in most cases for the estimates **X** and **s** (**sample** standard deviation) in samples taken from a normal distribution. In such a sample from a normal distribution, $\overline{X}$ was found to be 60 and $s = 15$. About 95% of the population values would be expected to lie between _____ and _____ .	30 90

3.16 If we randomly draw one element from a set of normally distributed observations, the probability that the element will be in the range $\mu \pm \sigma$ is .68; that is, 68% of all the population values are in this range. With a normal distribution where **X** = 1000 and **s** = 200, the approximate probability that a randomly drawn element will lie in the interval between 600 and 1400 is _____ .	.95

3.17 In situations where it is reasonable to assume an underlying normal distribution, the $\overline{X} \pm 2s$ (mean $\pm$ two standard deviations) interval is frequently used to indicate the central 95% of the distribution; that is, the end points approximate the 2.5 and 97.5 percentiles. Usually, a visual check of this assumption is made by simply noting whether or not the histogram is approximately symmetric and bell-shaped. For the data depicted below, the normal distribution assumption _____ (is, is not) reasonable.	is

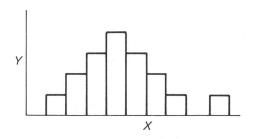

3.18 The utility of a calculated standard deviation is clear for samples from a normal distribution. This relationship between _____ and _____ allows for observations **whose distribution appears to be symmetric and bell-shaped; these relationships do not hold for other distributional shapes**.

$\overline{X}$,
s (or μ, σ)

If a sample mean $\overline{X}$ and sample standard deviation s were calculated from the data shown below, the interval $\overline{X} \pm 2s$ _____ (would, would not) contain about 95% of the values. The assumption of a normal distribution _____ (would, would not) be reasonable.

would not

would not
The distribution is clearly asymmetric.

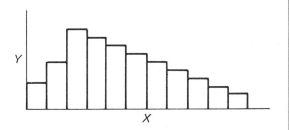

3.19 The dependence of the utility of a standard deviation upon the shape of the underlying distribution is often overlooked. Thus, $\overline{X} \pm 2s$ is frequently calculated and reported without an investigation of the shape of the distribution. Without knowledge of distributional shape, this interval _____ (does, does not) have a consistent physical interpretation.

does not

Distributions that are clearly asymmetric _____ (are, are not) well described by a mean and standard deviation.

are not

3.20 Because $\overline{X} \pm 2s$ estimates the 2.5 and 97.5 percentiles only in the case of Gaussian or normal distributions and **because most distributions encountered in nature are not normal, direct calculation of percentiles is usually preferred**. Valid use of percentiles does not depend on the underlying form or shape of the population distribution.

Part of the reason for the popular misuse of σ is the name "normal" distribution. **"Normal" is an adjective that applies to a mathematical model, and, in this sense, it has no biologic or clinical connotation.** Probably less confusion would result if the name _____ distribution were always used.

Gaussian

NORMAL LIMITS

3.21 It is important to describe as precisely as possible the population distribution of measurements of any variable that may be used in diagnosis or prognosis. This is true of anatomic measurements, physiologic measurements, or laboratory test values. The phrase **normal limits** is often used in this connection to refer to the upper and lower percentile points on a frequency distribution of such measurements made on "clinically normal" persons. It is crucial to specify which percentiles are implied when normal limits are referred to, since there is no particular set of percentile values that is always implied. Percentiles that have been suggested as a standard for normal limits are 5 and 95, 10 and 90, and, more recently, 2.5 and 97.5. Of course, the last may be simply estimated by the interval $\overline{X} \pm 2s$ when the distribution is _____.

Gaussian

3.22 In addition to being specific about the exact percentile values referred to by the phrase "normal limits," the nature of the population upon which the data are based should be made clear. Naturally, this population should have similar characteristics to those patients with whom it is to be compared.

During a routine examination, a patient's serum cholesterol value was reported to be within normal limits for his age-sex-race group. If the 5th–95th percentile values were used and are good estimates, the interpretation is that _____ This value lies between the 5th and 95th percentile values for healthy persons in his age-sex-race category.

3.23 It should be obvious that when two numbers are reported as the normal limits for some variable, it is important that two factors be specified for these limits to be meaningful:

1. _____ 1. The population (e.g., age, sex, race categories) from which these values were derived.

2. _____ 2. The exact percentile values (2.5–97.5, 10–90, or some other) implied for this distribution.

For an excellent discussion of normal limits in medicine, see Reference 3.

3.24 Some (e.g., Reference 3) have recommended that a consistency be established for the term **normal limits**, such as the 2.5–97.5 percentile cut points in a clinically healthy representative population. This laudable suggestion would thus include the central _____ % in the normal range. 95

NOTE: While the suggestion just presented in general is a good one, the clinical utility of a range depends on many factors (see Reference 15). The comparative distribution of values for both healthy and sick members of the appropriate population is the key factor. It is possible that, for some variables, whole populations are shifted so that defining "normal" becomes a broader issue. The comparison of total serum cholesterol values for Japanese and United States populations is such a situation. If we believe the Japanese values are normal, then a large portion of the entire United States adult population exceeds the normal limits.

The population-sample concept just described is fundamental to the understanding of statistical methods in medicine and public health. The next chapters elaborate on these methods using the idea of samples from defined populations.

Statistical Inference: Sampling Variation and Confidence Intervals

4

SAMPLING VARIATION

4.1 One of the reasons for the widespread usage of the standard deviation without sufficient consideration being given to the shape of the distribution involved is probably a misunderstanding of a mathematical theorem called the **central limit theorem**. This theorem has general utility in **making inferences about a population mean**, as opposed to **describing the variation of individual values** in a population. Apparently, these two concepts are sometimes confused.

Suppose we draw a random sample of size **n** from a population and designate the mean $\overline{X}_1$. We shall call the mean of a second sample of size **n**, $\overline{X}_2$. If we record the sample means for **many** (say k) **such samples** $(\overline{X}_1, \overline{X}_2, \overline{X}_3, \ldots, \overline{X}_k)$, we could construct a histogram for the frequency distribution of these $\overline{X}$s. The **central limit theorem** states that the **distribution of the means of such samples**, taken from a distribution of almost any shape, **tends to be a normal distribution**. The phrase "tends to be" means that a histogram of sample means looks more and more like a normal distribution as the number of observations in each sample increases.

If the population distribution is shaped like curve A, then the distribution of **individual observations**, Xs, contained in **one random sample** of **n** observations will be approximately shaped like _____ (A, B). However, the distribution of **means**, $\overline{X}$s, of **several such samples** will look about like _____ (A, B).

A

B

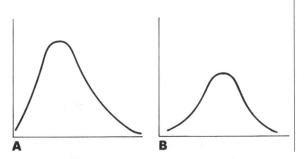

A　　　**B**

4.2 The central limit theorem is concerned with the variation between _____ _____ (individual observations, sample means).

sample means

4.3 The variance of the approximately normal distribution of sample means may be shown to be estimated by s^2/n, where s^2 is the variance calculated from the observations in **one** sample containing **n** observations. This variance of the distribution of sample means that may be calculated from the results of only one sample is designated by $s^2_{\bar{x}}$.

Variance of sample means $= s^2_{\bar{x}} = $ _____

s^2/n

4.4 The square root of this variance, $s_{\bar{x}} = \sqrt{s^2_{\bar{x}}} = \sqrt{(s^2/n)} = s/\sqrt{n}$, is sometimes called the **standard error of the mean**, or simply the **standard error**, **S.E.** This quantity is the standard deviation related to the normal distribution of sample means taken from a population. Thus, if **s** represents a sample standard deviation based on a sample of size **n**, then the expression $s/\sqrt{n}$ is called the _____ _____.

standard error

4.5 A particular set of **n** laboratory measurements may sometimes be considered as a **single random sample** of size **n** from a particular population. From these "sample" data, we can calculate the mean $\bar{X}$ and variance s^2.

   On the basis of the statements in sections 4.1 and 4.3, we can determine characteristics of the distribution of means of other samples of this size without actually taking further samples: We can say that the means of random samples from this population will tend to be distributed _____,

normally (or to
have a Gaussian
distribution)

and a standard deviation may be estimated from the data of our one sample to be _____ (symbols).

$s_{\bar{x}}$ or $\sqrt{(s^2/n)}$
or $s/\sqrt{n}$

4.6 The mean of the distribution of sample means may be shown to be (or to tend to be) equivalent to the mean of the parent distribution. Suppose, for example, that the parent population distribution of individual values, **X**, is shaped like curve A, below, with a mean as indicated by μ. The **means calculated from samples** randomly drawn from this population will distribute themselves normally about this population mean as shown by curve B. The two curves are superimposed in C.

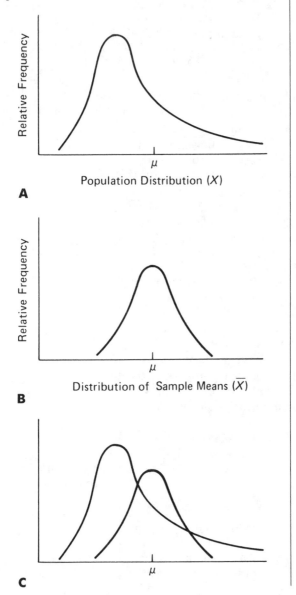

A

Population Distribution (*X*)

B

Distribution of Sample Means ($\overline{X}$)

C

Therefore, the distribution of means ($\overline{X}$s) of random samples taken from almost any distribution is approximately normal, with a mean equal to the unknown population mean μ and a standard deviation estimated by $s_{\overline{x}}$.

From our knowledge of normal distributions, we know that about _____% of the smple means will lie within $2s_{\overline{x}}$ units of μ, that is, within the interval $\mu \pm 2s_{\overline{x}}$.

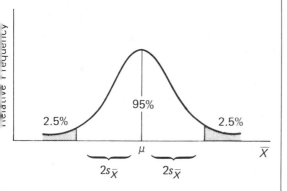

INTERPRETING CONFIDENCE INTERVALS FOR A POPULATION MEAN

4.7 In the case considered above, the parent population mean μ is unknown, but we do know that about 95% of the means ($\overline{X}$) of samples randomly drawn from a population will lie within $2s_{\overline{x}}$ units of μ, on one side or the other. It follows that the interval $\overline{X} \pm 2s_{\overline{x}}$ calculated for a particular $\overline{X}$ will "overlap" μ (as illustrated below) in 95% of the sample $\overline{X}$s.

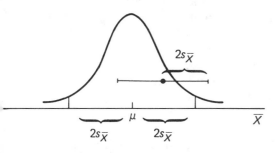

Therefore, we can say that there is a _____% chance that the interval $\overline{X} \pm 2s_{\overline{x}}$, calculated on the basis of only **one** randomly drawn sample, will "enclose" the true mean μ of the normal distribution of sample means, and this μ is equivalent to the population μ. We know that about ____% of such intervals will not enclose μ.

| | 5 |

4.8 Such an interval is called a **95% confidence interval for a mean**. The end points of the interval, $\overline{X} - 2s_{\overline{x}}$ and $\overline{X} + 2s_{\overline{x}}$, are called **confidence limits**. The interval $\overline{X} \pm 3s_{\overline{x}}$ would be a ____% confidence interval.

| | 99 |
| | (approximately) |

4.9 Since $s_{\overline{x}} = \sqrt{(s^2/n)}$, we can see that the more variable the parent population (and hence the larger s), the ____ (larger, smaller) will be $s_{\overline{x}}$. Also, the larger our sample size **n** is, the ____ (larger, smaller) $s_{\overline{x}}$ will be. The smaller $s_{\overline{x}}$ is, the ____ (wider, narrower) will be the confidence interval.

	larger
	smaller
	narrower

4.10 A confidence interval is, in a sense, a quantitative way of expressing how "confident" we are in our sample estimate ($\overline{X}$) of the population mean. It takes into account population variability and sample size. Sometimes $\overline{X}$ is reported along with its standard error, $s_{\overline{x}}$, since all confidence intervals may be formed by adding and subtracting the product of some constant times the standard error to the sample mean.

The interval $\overline{X} \pm s_{\overline{x}}$ may be termed a ____% confidence interval.

| | 68 |

4.11 The interval $\overline{X} \pm 2s_{\overline{x}}$ is approximately a 95% confidence interval only for fairly large sample sizes. For small sample sizes, the appropriate interval is $\overline{X} \pm ts_{\overline{x}}$, where **t** is found from a table of the **t**-distribution (given in most statistics texts). The **t**-distribution is further discussed later in this chapter and in much more detail in Reference 13. The value for **t** (with **n** $-$ 1 degrees of freedom) will be slightly larger than 2 (or 1.96), compensating for the fact that σ^2 is estimated from the sample data by s^2.

In general, $\overline{X} \pm ts_{\overline{x}}$ will be ____ (wider, narrower) than $\overline{X} \pm 2s_{\overline{x}}$.

| | wider |

4.12 The idea of a confidence interval for a mean, using s_x as calculated from measurements, depends upon the central limit theorem and does not depend upon the shape of the population distribution. This concept is concerned with the **distribution of sample means**, not with the **distribution of individual values** in the population.

As previously mentioned, there has been some confusion regarding the importance of the shape of the parent population distribution. The validity of the **confidence limit** concept does not depend on the assumption that the population is approximately normally distributed. The use of standard deviations to describe the variation of individual values _____ (does, does not) require this assumption.

does

4.13 One way to interpret confidence intervals (the end points are called **confidence limits**) for a population mean is as follows:

Suppose a large number of random samples, each of size **n**, were drawn from a single measurement population and the 95% confidence intervals $\overline{X} \pm t s_{\overline{x}}$ were independently calculated for each sample ($\overline{X}$ and s_x will vary from sample to sample). The means and confidence intervals for such samples could be displayed as follows:

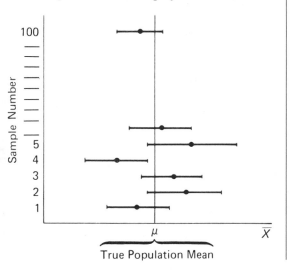

About 95% of intervals calculated in this way will enclose the population mean μ, and about ____% will not. | 5

4.14 The same interpretation applies to other percentage confidence intervals. If 99% confidence intervals for μ are calculated, then about _____% of such intervals will not enclose μ. | 1

4.15 Although the "long-term" percentages are known, we do not know whether or not any single confidence interval for a population mean encloses μ. Once a specific interval has been calculated, the true μ either is in the interval or is not. The probability statement associated with a 95% confidence interval is that 95% of the time, such intervals enclose μ. For any **specific** calculated interval, the probability that μ is enclosed is either _____ or _____ . | zero, one (either includes μ or does not) .

4.16 Confidence intervals provide a precise, objective (probabilistic) way of specifying how good a sample estimate $\overline{X}$ is. This method takes into account both sample size and population variability. Sample means based on large sample sizes are better than those based on a small number of observations. Thus, a large **n** leads to narrow confidence limits. Also, populations with relatively small variation will yield a small estimate of s^2, and the confidence intervals will be narrow. Of course, narrow confidence intervals imply that $\overline{X}$ is close to μ.

Estimates of two different population means and 95% confidence intervals were reported to be

A. 20 ± 8.5
B. 21 ± 3.5

If each were based on random samples, we would have greater faith in estimate _____ (A, B). | B

INTERPRETING CONFIDENCE INTERVALS FOR A POPULATION PERCENTAGE

4.17 In the previous section we considered the interpretation of confidence intervals for a population mean when we are dealing with continuous or measurement data. Next, we will consider the analogous situation with discrete data when we are interested in estimating a population percentage.

The first step is to define a **binomial population**. This is a **discrete** dichotomous population in which every element may be classified as belonging to either of two mutually exclusive classes. Examples are dead or alive, male or female, sick or well, and so on. Data obtained by sampling from a binomial population are usually used to construct a percentage, such as the percentage who are sick.

In a community health survey, for example, every person under 5 years of age was recorded as having been inoculated for measles or not. Such a two-class population is called a(n) _____ population.

binomial

4.18 For generality, the two classes of any binomial population are referred to as **successes** or **failures**. The percentage of successes in the population is equal to the probability of obtaining a success on a single "draw" (i.e., the random selection of one element). Thus, regarding the infinite population of results of tossing a fair coin, the probability of a head (success) on one toss is _____. Conceptually, this is equivalent to drawing one coin at random from a population of identical coins.

$\frac{1}{2}$

4.19 Suppose the results of a routine surgical procedure may be classified into two categories, improved or unimproved. Long experience has shown that the percentage of successes is 80%. One case drawn at random from this series has a probability of failure of _____.

.20, or 20%

4.20 A new surgical procedure tried on 20 patients shows 15 improved patients and 5 unimproved, a 75% success rate. Repetition of this procedure with other groups of 20 patients could, of course, result in different percentages of success. Such variation between sample percentages (i.e., those taken from a single population) is called **binomial sampling variation**. Knowledge of this variation allows the derivation of confidence intervals for a population percentage. However, the method is too complex to cover here. Many books (see References 4 and 13) contain tables from which such intervals corresponding to a sample percentage may easily be found.

From such tables, it can be seen that the 95% confidence interval for a sample result of 15/20 = 75% is 51% to 91%. Our best estimate of the population percentage is, of course, 75%. The interpretation of this interval is the same as that for confidence intervals for a population mean μ. About 95% of such intervals will enclose the true population percentage or success rate, and _____% will not.

5

NOTE: Whenever possible, confidence intervals should accompany point estimates. There are good point estimates and bad point estimates. Confidence intervals help us to differentiate between them using probability.

Statistical Inference: Tests for Statistical Significance

5

TESTS FOR STATISTICAL SIGNIFICANCE: COMPARING PERCENTAGES

5.1 Confidence intervals are usually a quantitative means of assessing the results of a **single** sample. Most biologic research, however, deals with comparative studies, rather than with the assessment of a single data set. In the simplest (and very common) situation, data from two samples are compared. In order to generalize usefully in such cases, the population-sample concept is employed. It is important to realize that **the results from an experiment may be regarded as a sample result from the hypothetical population of all possible experimental results if the experiment were repeated many times under "identical conditions."**

We shall first consider an experimental situation in which two percentages are to be compared. These data might have resulted from a comparison of the effects of two treatments or of a treatment and a control. The term **treatment** is used in its broadest sense, encompassing a variety of possible variables. We might consider two such percentages as having been obtained by sampling from _____ populations.   binomial

5.2 Because of binomial sampling variation, we would not expect two random samples from a single binomial population to be identical. If treatment group A yields 6 successes out of 10 and group B yields 5 out of 10, such a difference — 60% versus 50% — could well be attributed to _____ variation. If the observed difference were   (binomial) sampling
10 of 10 and 0 of 10, sampling variation _____ (would, would not) likely be considered   would not
the main factor in accounting for this difference.

5.3 When experimental results turn out to be extremely different or very much alike for two treatments, there is no decision problem regarding whether or not sampling variation could have easily accounted for the observed difference. When the observed differences are not so clear-cut, however, it is not obvious whether the apparent difference might reasonably be attributed to sampling variation. In such instances, it is useful to know the **probability** of this being the case.

One way of approaching this problem would be **to determine the distribution of the difference of two sample percentages if treatment effects are equal**. Of course, if they **are** equal, then the observed treatment difference must be entirely due to _____ or "chance."

sampling variation

5.4 The hypothesis that the treatment effects are equal is the **null hypothesis, H_0.** If H_0 is true ($A = B$), the two observed treatment percentages will _____ (often, seldom) differ to a great extent.

seldom

5.5 **Statistical significance tests or tests of hypothesis** have been devised **to assess the compatibility of a set of data with a null hypothesis H_0 of no treatment differences.** If we can determine that the results of an experiment are unlikely (have a low probability) to have occurred by sampling variation alone, we _____ (would, would not) be inclined to accept the null hypothesis. If the results are likely (have a high probability) to have occurred by sampling variation alone, we _____ (would, would not) be inclined to accept the null hypothesis.

would not

would

5.6 In fact, the use of statistical significance tests is based on formal decision-making rules concerning the compatibility of a set of data with the null hypothesis, **H_0,** of no treatment difference. Such rules lead to **accepting** or **rejecting** the null hypothesis. If our scheme leads to a rejection of **H_0,** then we would conclude that something other than _____ _____ must be contributing to our observed treatment differences; this would be a true _____ _____.

sampling variation

treatment difference

5.7 To put such decision-making on a probabilistic basis in order to compare two percentages, it would be useful to obtain the distribution of treatment differences under (i.e., given) the null hypothesis. However, in this situation, it has proved to be more convenient to compute a single statistic, the 2×2 χ^2 _____ **test**. Although there are other significance tests useful for this comparison, the χ^2 is the most popular. The word **statistic**, in the sense used here, means a value calculated from the observed data that summarizes (in some meaningful way) the evidence against H_0.

(chi-square)

The distribution of this calculated statistic χ^2 under H_0 has been determined. Small observed treatment differences lead to **small** values of χ^2; large observed differences result in **large** values of χ^2. Thus, if we were to use the computed χ^2 value as a basis for deciding in favor of or against H_0, we would decide against H_0 if χ^2 were too _____ (large, small).

large

5.8 As previously mentioned, the two classes of a binomial population are usually called **successes** or **failures**. In the comparison of any two percentages, a 2×2 table may be formed as follows:

	Number of		
Group	Successes	Failures	Total
A	**a**	**b**	**a + b**
B	**c**	**d**	**c + d**
Total	**a + c**	**b + d**	**n**

The success rate for A is _____ [**a**/(**a** + **c**), **a**/(**a** + **b**)].

a/(a + b)

5.9 For any such 2×2 table constructed as above, the χ^2 value may be computed by the following formula*:

$$\chi^2 = \frac{(|\mathbf{ad} - \mathbf{bc}| - \mathbf{n}/2)^2 \cdot \mathbf{n}}{(\mathbf{a} + \mathbf{c})(\mathbf{b} + \mathbf{d})(\mathbf{a} + \mathbf{b})(\mathbf{c} + \mathbf{d})}$$

Inspection of this formula shows that a calculated χ^2 _____ (can, cannot) be negative.

cannot

*The vertical lines indicate absolute value.

5.10 The formal decision-making procedure in this case may be derived as follows. Suppose two random samples are drawn from a single binomial population (which is the case when H_0 is true) and the value of χ^2 is calculated for these two sample percentages. Suppose also that this procedure is repeated a large number of times and the distribution of calculated χ^2s is found. Because of sampling variation, there would be a few sample pairs with widely different percentages, and in each such case, a _____ (large, small) χ^2 would be calculated from the corresponding 2×2 tables.

large

5.11 The distribution of such calculated χ^2 values for nearly any binomial population is shown below:

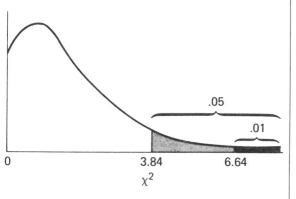

Only 5% of such χ^2s will exceed 3.84, so the probability of a χ^2 being less than 3.84 is about _____ when H_0 is true.

.95, or 95%

5.12 The above distribution is called a χ^2 **distribution with one degree of freedom** (Reference 1, 2, or 13). It is applicable to the comparison of two percentages in a 2×2 table. It is important to recall that this distribution was derived by considering all possible values when two samples were taken from a **single** _____ **population**, that is, given that the _____ hypothesis was true. If the probability of a success with treatment A, P_A, equals that for treatment B, P_B, then $P_A = P_B = P$ (a single-population success rate).

binomial

null

5.13 Large values of χ^2 are possible under H_0, but they occur with a low, specifically determined probability. Five percent of the time, χ^2 will exceed 3.84; 1% of the time, it will exceed 6.64. Thus, we may use the calculated χ^2 value as a formal basis for accepting or rejecting H_0, that is, for assessing the compatibility of the data with H_0. If H_0 were true and we rejected H_0 when $\chi^2 > 6.64$, we would be in error _____ % of the time. If we rejected H_0 when $\chi^2 > 3.84$, we would be in error _____ % of the time when H_0 is true.

1

5

5.14 Using the extreme χ^2 values of 5% (or .05) and 1% (or .01) as the criteria for accepting or rejecting H_0 is called **testing the null hypothesis**. If the calculated χ^2 **equals** or **exceeds** 3.84, we say H_0 **is rejected at the 5% level**, or **.05 level of significance**. If our χ^2 is **less** than 3.84, we **accept H_0** at the _____ level.

5%, or .05

5.15 In a comparison of two treatment percentages, if H_0 is rejected at a stated level of significance, we state that there is a **statistically significant difference** between the two treatments. Thus, if the corresponding χ^2 value is greater than 3.84, we say there is a statistically significant difference between treatment effects at the _____ level.

.05

5.16 If $\chi^2 < 3.84$, we say that no statistically significant difference was found. Therefore, we conclude that the two treatments are _____ (different, equal), and, accordingly, we _____ (accept, reject) H_0. This does not constitute proof, however, as will be discussed later.

equal

accept

5.17 Suppose two drug treatments are being compared using success rates. The rate for drug A is 6/10 = 60%; the rate for drug B is 2/10 = 20%. The appropriate 2×2 table is

Drug	Number of patients Improved	Unimproved	Total
A	6	4	10
B	2	8	10
Total	8	12	20

From the χ^2 formula, the computed χ^2 value is 1.88 in this case. Therefore, we would _____ (accept, reject) H_0 at the _____ % level of significance.

accept
5

5.18 Another way of stating the results of the above χ^2 test is that **the probability of obtaining a difference in treatment percentages as large as, or larger than, that observed** (6/10 versus 2/10) **is greater than** _____ **if** H_0 **(no real treatment difference) is true.** Alternatively, we may write **p** > .05; i.e., the above probability is greater than .05. This is equivalent to saying that the difference _____ (is, is not) statistically significant at the 5% level.

.05

is not

5.19 Accepting H_0 in this case is equivalent to concluding that the two treatments are _____ (equal, unequal) in their ability to cause improvement. We know they **may** be unequal, but the difference could easily (**p** > .05) be explained by random sampling variation, or "chance," alone.

equal

5.20 If our calculated χ^2 value exceeded 3.84 in the above example, we would state that the observed treatment difference _____ (was, was not) statistically significant at the .05 (or 5%) level of significance. Alternatively, we may write **p** < .05, indicating that H_0 was rejected at this level of significance. Another way of stating this result is that **the probability of obtaining a difference in treatment percentage as large as, or larger than, that observed is** _____ (greater, less) **than .05 if** H_0 **(no real treatment difference) is true.**

was

less

5.21 Rejecting H_0 in this case is equivalent to concluding that the two treatments are in fact _____ (equal, unequal) in their ability to cause improvement. We know that they **may** be equal, but such an observed difference could not easily ($p < .05$) be explained by random sampling variation alone.

unequal

5.22 Another significance test that is often used to compare two percentages, particularly in experiments (see Chap. 7), is **Fisher's Exact test** (see References 1 and 4). This test is also sometimes used instead of the χ^2 test when the sample sizes are very small, say less than 10. If we wished to compare 20/30 versus 10/20, we would probably use the _____ (Fisher's Exact, χ^2) test.

χ^2

NOTE: For a detailed discussion of the general formulation of the χ^2 test, its extension to the comparison of more than two percentages or rates, and other related tests, see References 1, 2, 4, and 8, or other basic statistical methods texts.

TESTS FOR STATISTICAL SIGNIFICANCE: MEASUREMENTS

5.23 A large number of statistical significance tests are available for testing null hypotheses when measurements, not percentages, are involved. We shall consider only two such tests, one for **paired measurements** and one for **unpaired measurements**. Experimental designs employing **matched pairs** of experimental subjects and designs using two groups of **unpaired subjects** are the two most frequently employed methods for comparing two treatments.

If the set of available experimental subjects can logically be grouped into homogeneous subsets of size two (by using littermates or other such means), we could take advantage of this situation and use _____ (matched pairs, unpaired subjects) for our experimental design.

matched pairs

5.24 One type of paired-measurement design in clinical studies is used when the patient "serves as his own control." By this, it is meant that each subject is treated at two different time periods, each time with one treatment, A or B. If one of the treatments is a placebo (see Chap. 7), then he does serve as his own _____.

control

5.25 Suppose in a paired drug study involving antihypertensive drugs A and B, the following measurements in average reduction of diastolic blood pressure (DBP) were made:

	DBP (mmHg) reduction		
Patient no.	Drug A	Drug B	Difference $(A - B)$ d_i
1	15	20	-5
2	30	20	10
3	10	4	6
4	25	20	5
5	12	10	2

The evidence regarding a difference in average drug effects is summarized in the set of differences, d_i, shown in the right-hand column. If the null hypothesis is true (that is, drug A = drug B), then the true mean difference μ_d = _____. This value, μ_d, refers to the mean difference in d_is in the entire patient population under consideration. These five d_is might be regarded as a sample from this population.

0

5.26 Thus, the appropriate null hypothesis in paired studies may be written as $\mu_d = 0$. As was the case with percentages (in fact, with all significance tests), we want to assess formally the compatibility of our data (d_is) with the null hypothesis $H_0 : \mu_d = 0$. To do this, we will first need to consider the variation between pairs, which may be summarized by a histogram of the d_is. Construct such a histogram (below) for the data in section 5.25:

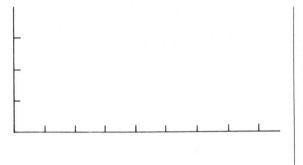

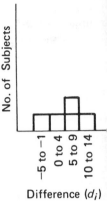

No. of Subjects

−5 to −1 · 0 to 4 · 5 to 9 · 10 to 14

Difference (d_i)

5.27 The mean difference in the above case, $\bar{d} = +3.6$ mmHg, must be viewed against the between-pair variability as illustrated by the histogram of differences, d_is. For example, the same mean $\bar{d} = 3.6$, would be considered as stronger evidence against H_0 if this variability were _____ (less, greater) than that shown in this example.

less

5.28 The most frequently used statistical test of the null hypothesis $\mu_d = 0$ is known as the **Student's paired** t-test. This is also known as simply the **paired** or the **one-sample** (of d_is in this case) t-test. For validity — that is, to ensure that the probability statements are exact — the d_is must follow a Gaussian distribution. However, for most situations met in practice, paired t-test results will be approximately correct even if this assumption is not fulfilled. Therefore, for validity, the normal-distribution assumption _____ (is, is not) crucial for paired t-test validity.

is not

5.29 The statistic calculated for the paired t-test on **n** pairs is given by the formula

$$t = \frac{\bar{d}}{s_{\bar{d}}}$$

where $\bar{d}$ is the "sample mean" (i.e., the observed average difference) and $s_{\bar{d}}$ is the standard error computed for the d_is. The distribution of the

-statistic **is known** under the assumption that H_0 is true. Under H_0, it is unlikely that **d** (and hence) would be very _____ (close to, far from) zero. | far from

.30 It can be seen that this statistic takes the between-pair variability into account in a logical manner. Since

$$s_{\bar{d}} = \frac{s_d}{\sqrt{n}}$$

he larger the variation among the **d**s (as measured by s_d), the _____ (larger, smaller) $|t|$ is ikely to be. | smaller

.31 The distribution curve for **t** when H_0 is true s shaped much like a Gaussian distribution, as shown below:

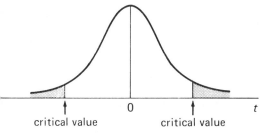

critical value   0   critical value   t

Unlike the χ^2 statistic, a **t**-value may be either positive or negative. It has a mean at zero (perfect agreement with H_0). Each of the shaded areas in the tails of the distribution contains 2.5% of the total probability; that is, the probability that a random sample of **d**s, under H_0, would yield a calculated **t** in one of these shaded areas is _____. The probability that a calculated **t** would all in the unshaded center is _____. | .05 .95

.32 The **t**-values — called the **.05 critical values** — that mark off the extreme 5% (2.5% in each tail) in the **t**-distribution vary with the sample size **n** (the number of pairs). This is unlike the χ^2 critical values 3.84 (.05) and 6.64 (.01), which are independent of sample sizes. The critical values of **t** may be found from a **t**-table (given in statistics texts) under **n** $-$ 1, the "degrees of freedom." If

$n = 5$, for example, the critical t-values will be ± 2.776 at the 5% probability level. This means that 95% of all calculated ts, based on $n = 5$ and under H_0, will lie between -2.776 and $+2.776$. If a calculated t (based on five pairs) lies outside this interval, we know the probability of this event, given H_0 is true, is less than _____. .05

5.33 The critical values for t that exclude only 0.5% in each trial are called the **1% critical values**. For $n = 5$ pairs, these are $t = \pm 4.604$. If there is no treatment difference, the probability that the absolute value of a calculated t will exceed 4.604 is _____. .01

5.34 These critical values for t may be used as the decision criterion for accepting or rejecting H_0, just as was the case for χ^2 in the comparison of two percentages. If the absolute value of the calculated t exceeds the critical value, then we would _____ (accept, reject) H_0. If the $|t|$-value is small (i.e., between the two critical values), we would _____ (accept, reject) H_0. This is intuitively reasonable, since small values of $|t|$ would often occur under H_0. reject

accept

5.35 If the absolute value of a calculated t-value is less than the critical t-value so that H_0 is accepted, we then say that there is **no statistically significant treatment difference**. In the example of section 5.25 where the five differences were -5, 10, 6, 5, and 2 mmHg, $\bar{d} = 3.6$ and the calculated t is $+1.439$. Since the 5% critical t-value is $+2.776$, we would conclude that the difference between the response to drugs A and B _____ (was, was not) statistically significant at the .05 level. was not

5.36 Suppose drugs A and B are being compared in an **unpaired** study having 6 different patients per group, and the following results based on the 12 patients are obtained:

Measurement (changes from patient's baseline)	
Drug A	Drug B
8	10
−2	12
6	4
6	11
5	9
1	8
$\overline{X}_A = 4.0$	$\overline{X}_B = 9.0$

In this case, there is no pairing, and the appropriate null hypothesis may be stated as H_0: $\mu_A = \mu_B$; that is, the average treatment effects are equal for the relevant patient populations.

The evidence concerning H_0 may be expressed as the observed difference between the mean treatment effects (5.0 in this case), but we must again take the between-subject variation into account. Thus, the evidence contrary to H_0 would be considerably stronger with an observed difference of 5.0 units if the within-group variability were _____ (less, more) than that shown here. | less

5.37 The most frequently employed statistical test of the null hypothesis $\mu_A = \mu_B$ with unpaired data is the **unpaired (Student's)** t-**test** or the **two-sample** t-**test**. The formula is

$$t = \frac{\overline{X}_A - \overline{X}_B}{s_{(\overline{X}_A - \overline{X}_B)}}$$

where the numerator is the difference between the means for the two treatment groups and the denominator is the measure of variation of mean differences, called the **standard error of differences**. For a discussion of this test (including computational details), see Reference 13 or any other standard statistics text.

This t-statistic is calculated and compared with a t-distribution with $n_A + n_B - 2$ degrees of freedom, where n_A and n_B are the number of subjects (sample sizes) in each group. For the data in section 5.36 the appropriate degrees of freedom would be _____ . The two sample sizes need not be the same.

10

5.38 The interpretation of a calculated **t** for unpaired measurements is exactly the same as the calculated **t** for paired measurements. Large absolute values of **t** are required in order to _____ (accept, reject) H_0.

reject

5.39 For the example shown in section 5.36 the computed **t**-value is -2.61. From tables of the **t**-distribution, it can be seen that the 5% critical **t**-value for 10 degrees of freedom is 2.228. Since $|-2.61| > 2.228$, we would _____ (accept, reject) the null hypothesis of equal treatment effects. Thus, the observed mean difference is large enough — relative to the between-subject variability — to conclude that there is a real treatment difference. If there is no real treatment difference, the probability of obtaining an observed **t** as large as that found is _____ .

reject

$\leq .05$

5.40 For validity, the two-sample (unpaired) **t**-test requires that the observations in each group be normally distributed and that the variances, σ^2, be the same. When these assumptions are clearly violated, other statistical tests may be used (see Reference 13).

The assumption of equal variances in each group _____ (is, is not) probably satisfied for the data in section 5.36.

is (the ranges are 10 and 8, the s^2s are 14 and 8)

NOTE: A technical point should be made concerning the direct link between confidence intervals and statistical significance tests. A confidence interval for a mean difference that does not include the value zero indicates rejection of the null hypothesis of equal group effects or group means. Inclusion of zero implies acceptance of H_0. See Reference 1 or 13 for examples.

For very large sample sizes (e.g., **ns** of ≥ 100)

t-tests become practically equivalent to a test called the **normal or z-test**. Here t-values become z-values. The interpretation is the same as with a t-test.

TWO KINDS OF DECISION ERRORS

5.41   The traditional statistical decision-making process, so widely adopted in areas of science such as medicine where variation is an important phenomenon, can be represented in a table similar to the decision table regarding diagnostic tests presented in Chapter 1 (section 1.12).

We may dichotomize the decision-making about H_0 using the data at hand in a statistical test, as shown in the following table:

Decision based on statistical test	*True situation*	
	H_0 *is true*	H_0 *is false*
Accept H_0	No error	Type II error
Reject H_0	Type I error	No error

From this table we can see that rejecting H_0 (on the basis of our test), when it is in fact true, is called a type _____ error.   I

5.42   The probability of making this error using our test is called α (alpha). This is the general term for our selected significance level or _____   p value. Thus, in this context, we can state that our selected α- or **p**-value is our predetermined (before seeing the data) probability of making a type _____ error.   I

5.43   If we erroneously accept H_0, the error is called a _____ error. The probability of making   type II this error is termed β (beta). **Clearly this probability is dependent on how much the true situation deviates from H_0.** An examination of the construction of common statistical tests will reveal this fact. The greater the true hypothesis deviates from H_0, the _____ (larger, smaller) β will be. | smaller

5.44 The quantity, one minus the probability of a type II error, or $1 - \beta$, is the probability of making a(n) _____ (correct, incorrect) decision | correct
under the H_0 false situation. This probability is called the **power** of a statistical test such as a χ^2 or **t**-test. Clearly the power _____ (increases, | increases
decreases) the more the true hypothesis deviates from H_0.

NOTE: The preceding statements about power assume that all other factors remain constant. Actually, the power of a test is a function not only of the true deviation from H_0 but also the choice of α (usually .05 or .01), the variance involved, and the sample size. Coverage of these complex relationships is beyond the scope of this book. The concept of power is extremely important in the interpretation of statistical tests. Methods texts such as References 1, 2, 8, and 13 provide a fuller discussion.

Testing of a null hypothesis, H_0, means making a decision about the truth or falseness of H_0 based solely on the data at hand. Methods for incorporating and quantifying prior information into this process belong in the arena of **Bayesian statistical inference**. The interpretation presented here is restricted to the more widely applied and accepted approach to statistical inference in the biomedical sciences.

STATISTICAL AND PRACTICAL SIGNIFICANCE

5.45 There are a large number of statistical tests of null hypotheses for various research situations. We have only considered three of the most frequently employed tests: the _____ test for comparing two percentages, the _____ **t**-test, and the _____ **t**-test. The **interpretation** of **all** such significance tests, however, is the same. A statistic | $2 \times 2 \ \chi^2$
paired (or one-sample), unpaired (or two-sample)

(i.e., χ^2, **t**, or other) that summarizes the evidence against the null hypothesis, H_0, is calculated and compared to the distribution of such statistics **if the null hypothesis is true**. If the calculated statistic is judged to be too unlikely under H_0, then H_0 is _____ (accepted, rejected). Otherwise, it is _____ (accepted, rejected) at the stated _____ level.

rejected
accepted
significance

5.46 In summary, obtaining a **statistically significant difference** (say, $p < .05$) between two treatments indicates that **differences as large as, or larger than, those observed may occur with "too small" a probability under** H_0 **to be reasonably** (in our judgment) **attributed solely to chance**. We can specify what we mean by "too small"; in this case, it is _____. This is the probability to which a **p**-value refers (i.e., $p < .05$, $p < .01$, $p < .001$, and so on).

.05

5.47 Conversely, we have stated that obtaining a test result that indicates **no statistically significant difference** between two treatments implies that **differences as large as, or larger than, those observed may occur under** H_0 **with "too large" a probability for us to reasonably rule out the null hypothesis**. We can specify with probability values what we mean by "too large." If we are dealing with the .01 significance level, we mean that **p** is _____ ($<$, $>$), .01, or 1%.

$>$

5.48 Thus, a difference observed between the results of two treatments that is judged to be statistically significant ($p < .05$) means _____

(phrase in your own words)

differences as large as, or larger than, that observed would occur by chance alone **less than** 5% of the time

Conversely, a difference observed between the results of two treatments that is judged not to be statistically significant ($p > .05$) means _____

(phrase in your own words)

differences as large as, or larger than, that

5.49 A researcher reports in a medical journal article that he has compared the mean healing times of two treatments for minor burns in children at a large clinic. He states that treatments were randomly allocated to "similarly burned" children as they presented for care and that the study terminated after 20 children had been enrolled and followed in each group. The mean difference in healing times — 2.5 days in favor of treatment A — was statistically significant at the .05 significance level. The statistical test used was probably the _____. Rephrase this result in terms of a probability statement:

_____
_____
_____
_____

In this case, one _____ (accepts, rejects) the H_0 of no treatment difference.

observed would occur by chance alone **greater than** 5% of the time

two-sample **t**-test

The probability of observing a difference as large as or larger than 2.5 days is less than .05 if there is no real healing time difference (**p** $<$ .05).

rejects

5.50 A research report states that 100 male cigarette smokers and 100 male nonsmokers were drawn at random from a factory population. The percentage having missed at least 1 day of work due to "illness" during the last 6 months was tabulated. The difference (20% for smokers versus 12% for nonsmokers) was found not to be statistically significant at the 5% level. The appropriate statistical test is the _____.

Interpret this result: _____
_____
_____
_____

$2 \times 2 \; \chi^2$

A difference in percentages as large as or larger than that found (8%) could easily

($p > .05$) be explained by chance alone (two random samples from one binomial population).

5.51 Obviously, a null hypothesis is neither proved nor disproved by the statistical tests described here. "Acceptance" and "rejection" of H_0 are merely terms that relate to specific probability statements as described. We do not know whether or not H_0 is true (if we did, there would be no problem); however, the probabilities referred to are all computed under the hypothetical condition that H_0 is _____ (true, false).

true

5.52 In any statistical test, the probability statements relating to our acceptance or rejection of H_0 depend entirely on the randomization process for validity. If we are comparing samples from two populations, they should be _____ samples. If we are assigning treatments to experimental subjects, **these assignments should be made by a random process** (tables of random numbers are available in statistics texts for this purpose). Otherwise, the probability statements are inappropriate.

random

5.53 The choice of .01 and .05 as the usual significance levels is conventional but arbitrary. Obviously, a result that is statistically significant at the .01 level _____ (would, would not) also be statistically significant at the .05 level. Other levels may also be selected (References 1, 2, and 13 contain detailed discussions on the choice of significance levels).

would

5.54 It must be kept in mind that **statistical significance** is **not** equivalent to **practical** or **clinical significance**. Statistical significance pertains only to the **existence** of a difference, not its **magnitude**. To make a judgment about the practical significance of a finding, the estimated differences (magnitude) must be considered in the light of all accumulated evidence of which the investigator has knowledge.

Suppose, for example, that the difference in dissolving time in water for two types of analgesic pills was found to be 3.5 seconds. Even if this difference turned out to be statistically significant, and indeed real, it could hardly be considered _____ significant.

practically or clinically

5.55 Conversely, of course, real and important differences may be missed even when the data do not yield a statistically significant difference at a conventional significance level (usually .05). Not obtaining a statistically significant difference does not prove that no real difference exists; it only shows that the observed difference **could** easily be explained by chance alone. As we have indicated, small sample sizes, large population variability, and small real differences are all factors that militate against differentiating by a statistical test a real difference from random processes. Thus, if after calculating a valid statistical test, no statistically significant treatment difference is found, it _____ (is, is not) possible that a real and important treatment difference exists. It is possible that the test had very low power to detect important true differences!

is

5.56 Large sample sizes may be required to detect real, but small, treatment differences. Also, the smaller the variation is between subjects that are treated alike, the _____ (more, less) likely is the detection of a real difference. Therefore, homogeneous subject groups are generally desirable.

more

5.57 Small sample sizes that yield a nonstatistically significant result simply may not have enough _____ to detect real and important differences between two treatments or groups.

power

NOTE: There is a tendency to overemphasize statistical significance tests in the evaluation of data. One problem is that many researchers use such

Statistical Inference: Tests

tests and quote the corresponding **p**-values without a precise understanding of their interpretation. Another major factor is that with measurement data, most significance tests deal only with the difference between **mean** treatment effects. Especially when human subjects are involved, the full distribution of treatment effects and differences, not just means, should be carefully considered.

Relatedly, there should often be more concern with estimating the **amount** of mean difference with the corresponding confidence interval, rather than restricting the analysis to a simple test of statistical significance. It is also good scientific practice to accompany point estimates of differences with confidence intervals (when possible), whether or not a statistical test was employed.

Linear Regression and Correlation

6

LINEAR REGRESSION

So far, we have been mainly concerned with a single observation made on each element of a sample. We will now consider the case where two measurements are made on each element, that is, where the sample consists of pairs of values, one for each of two continuous variables **X** and **Y**.

This situation can arise in both experimental and nonexperimental contexts. In an **experiment**, we may choose (or control) values of the **independent variable X** and observe the **dependent variable Y**. In a **nonexperimental study** or survey, measurements are made on variables **X** and **Y** for each member of the sample. In this case, there is no control of levels and no true independent variable (although one is usually labeled as such).

6.1 Suppose the following responses, **Y**, were observed in an experiment involving five animals, each subjected to a different level of variable **X**. Construct a graphic representation by plotting the data.

Animal number	X	Y
1	−1	5
2	0	11
3	1	26
4	2	27
5	4	43

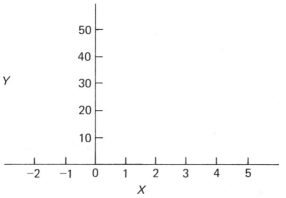

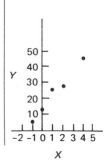

6.2 From the graph, there definitely appears to be an association between the two variables. In such cases, it is often desirable to employ a mathematical description of such an **X-Y** relationship. The simplest such expression is the general equation for a straight line:

$$Y = A + BX$$

This **linear model** or **simple linear regression** formula contains two parameters: the **intercept** A and the **slope** B. The intercept **A** is the value of **Y** when **X** = _____. The slope **B** is the **amount of change of** Y **for a unit change in** X.

0

6.3 In the regression equation **Y** = 5 + 2**X**, the numerical value for the intercept is _____ and the slope is _____. This equation is said to have a _____ (positive, negative) slope; that is, when **X** increases, **Y**_____ (increases, decreases).

5
2
positive
increases

6.4 In the equation **Y** = 8 − 0.5**X**, the dependent variable **Y** decreases _____ unit(s) as **X** increases one unit. Thus, the value of the slope is _____.

½

− 0.5

6.5 The true values of **A** and **B** are never known but must be estimated from the data. One way is to formulate "eyeball" estimates from the graph. However, because of between-person variability in judgment, this is usually not very satisfactory.

The most commonly employed procedure for calculating estimates of **A** and **B** is known as the **method of least squares**. If the data may be represented by the points below,

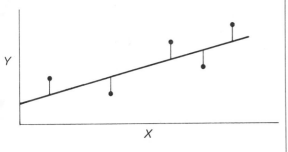

the **least-squares**, or "best-fitting," straight line shown is the line that minimizes the sums of the squares of the vertical deviations between the points and the line (computational details may be found in References 1, 2, or 13). Here, the slope is _____ (positive, negative).

positive

6.6 Least-squares formulas allow calculation of the best estimates of **A** and **B** from a set of **X-Y** measurements, whether or not the data are experimental. These estimates are conventionally labeled **a** and **b**, and the fitted regression equation is

$$Y = a + bX$$

Determine what the value of **a** must be for the following data. _____

zero

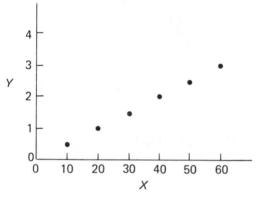

Of course, if a line must go through the origin, the underlying regression equation reduces to _____ (symbols).

Y = **BX** (Since **A** = 0, it may be deleted.)

6.7 One of the major uses of a regression line is to provide a formula for predicting **Y**, given a value for **X**. A predicted **Y** is designated **Ŷ** ("Y-hat"), and the estimated line, since it is usually used for prediction, is often written

$$\hat{Y} = a + bX$$

Given the estimated regression line **Y** = −10 + 3.25**X**, the predicted value **Ŷ**, if **X** = 10, is _____. The estimated slope is _____.

22.5, 3.25

In the absence of theoretical reasons to do otherwise, a linear model is usually assumed or "fitted" to data when the relationship between two continuous variables is investigated. $Y = A + BX$ is the simplest mathematical function from a wide variety of possible functions $Y = f(X)$ that could relate X and Y. It is **always** a good idea to plot the data on an X-Y graph to obtain a picture of the form of the X-Y relationship as a means of checking the plausibility of the model.

Make such a plot of the following data. By inspection, it appears that the relationship _____ (is, is not) linear.

is not

X	Y
.0	10
.0	39
.5	44
.0	47
.5	48
.0	46
.5	52
.0	38
.5	24
.0	24

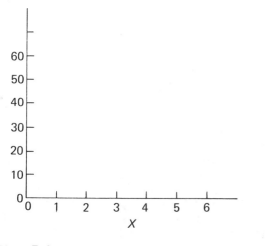

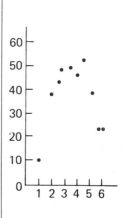

NOTE: Relationships, such as the one above, that could not be well represented by a straight line are called **nonlinear** or **curvilinear**. Many types of mathematical models are available for expressing the **X-Y** relationship in these cases but will not be considered here. Always construct a plot of the data to gain a visual idea of the relationship!

6.9 It is often of interest to investigate the question of whether or not one can predict **Y** from **X** by the regression line; that is, is there a linear relationship? This may be phrased in terms of the validity of a null hypothesis, $H_0: B = 0$. If H_0 is true, we have $Y = A$, which is the equation for a _____ line. This situation would imply that knowing **X** is of _____ (no, some) value in predicting **Y**, that is, that there is no relationship between **X** and **Y**.

flat (horizontal)
no

6.10 Sketch the regression line below for the case $A = 10$, $B = 0$.

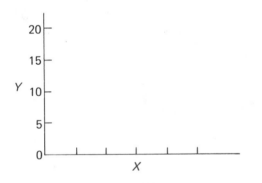

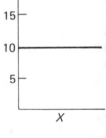

$Y = 10 + (0)X$
$Y = 10$

6.11 A type of **t**-test is used to test the hypothesis that the slope **B** is zero. The t-distribution is known if H_0 is true, and the absolute value of the computed **t** must exceed the labeled critical values in order to "reject" H_0 and conclude that **B** _____ ($=$, $\neq$) 0 (computational details and examples are found in Reference 13).

$\neq$

6.12 The interpretation of a test of the null hypothesis $H_0: B = 0$ is the same as that previously described for other statistical significance tests; that is, a statistically significant result at the .05 level implies that the probability of obtaining a

Linear Regression and Correlation

calculated **b** so different from zero, when **B** does equal zero, is less than _____.

.05, or 5%

6.13 Given the following calculated regression line (shown below), what is the predicted heart rate (beats/min) for a person who will receive isoproterenol at the rate of 0.10 mg/kg/min? (Note that neither scale begins with zero.) _____

133

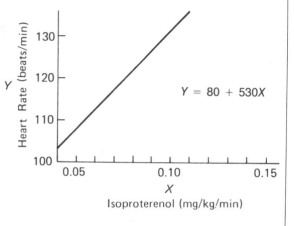

$Y = 80 + 530X$

LINEAR CORRELATION

6.14 When paired measurements are obtained from nonexperimental data (often called a **survey**), an underlying mathematical model called the **bivariate normal distribution** is often assumed. For practical purposes, this is equivalent to assuming that both continuous variables (**X** and **Y**) are normally distributed. One of the parameters of the bivariate normal distribution is called the **linear correlation coefficient** ρ (rho). This parameter is a **measure of the strength of the linear association of** X **and** Y.

The limits on ρ are −1 to +1; any other number in between is possible. A positive value for ρ indicates a **positive** linear relationship; that is, as **X** increases, **Y** increases. A **negative** ρ indicates a _____ linear relationship; that is, as **X** increases, **Y** _____. If ρ = 0, this indicates that **X** and **Y** are unrelated; that is, as **X** increases, **Y** _____.

negative
decreases

does not change

6.15 If $\rho = +1$ or -1, this means that the **X-Y** relationship exactly follows a straight line; these are the only such cases in which this occurs. Of course, when $\rho = -1$, the line has a _____ negative (positive, negative) slope.

6.16 Some examples of the special possibilities noted above are graphically displayed below. Such **X-Y** plots in nonexperimental situations are called **scatter diagrams** or **scattergrams**.

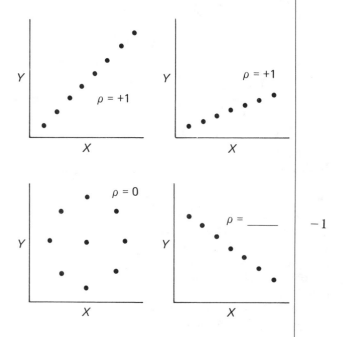

-1

6.17 Since ρ is a measure of the strength of linear association between two variables, it is often desirable to estimate it from a random sample from the population. This **estimated** correlation coefficient is called **r**. Formulas for **r** may be found in nearly any statistics text. Like ρ, **r** must lie between -1 and $+1$. Some examples are shown in the following scattergrams:

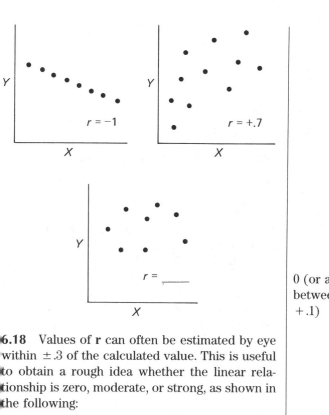

0 (or at least between −.1 and +.1)

6.18 Values of **r** can often be estimated by eye within ±.3 of the calculated value. This is useful to obtain a rough idea whether the linear relationship is zero, moderate, or strong, as shown in the following:

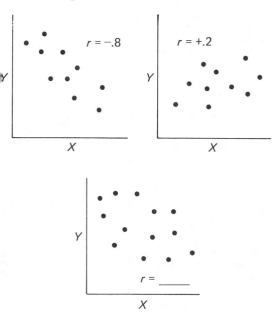

−.5 (between −.2 and −.8)

6.19 A study based on a random sample of 50 adult male residents, ages 20 to 29, from a large defined community, measured the high-density lipoprotein (mmol/L), or HDL, and a measure of relative weight in terms of percent overweight. The calculated correlation coefficient for these two variables was -0.40. The implication from these data is that the more overweight a young man is, the _____ (higher, lower) his HDL is likely to be.

lower

6.20 Because a value obtained for **r** is an estimate of a population parameter ρ, it may only be interpreted in the manner discussed if it is based on a random sample from a population. Therefore, a correlation coefficient is **not** of much use with experimental data where the values of the independent variable are fixed and chosen by the experimenter. For this reason, a correlation coefficient _____ (is, is not) usually calculated for experimental data.

is not

6.21 Because, strictly speaking, **causal inferences cannot be drawn without experimentation**,* the fact that two variables are found to be correlated in nature means only that they are related. It does not imply a causal relationship; that is, **correlation $\neq$ causation**.

A third variable (say **Z**), known or unknown, may be causally related to both **X** and **Y**. Changes in the value of **Z** in either direction may cause changes in **X** and **Y**. For example, a positive correlation may be found between population cigarette-smoking rates, **X**, and incidence of "athlete's foot," **Y**. If a "constitutional factor," **Z**, causes people to choose to smoke and to be more active, and if physical activity is related to this disease, then we would say that **Z** is causally related to **X** and **Y**, but that **X** and **Y** are not necessarily _____ related.

causally

6.22 The regression equation itself is an estimate of the linear relationship between two vari-

*With rare exceptions. It would be instructive to construct some examples. This topic is considered in Chapters 7 and 8.

ables. The correlation coefficient **r** calculated from a sample is a measure of the **strength** of the **linear** association. The variables shown below are strongly related, but their **linear** relationship is weak over **the full range of** X. A calculated **r** in this case would be about _____.

zero

6.23 One hypothesis that is often of interest in analyzing nonexperimental data is H_0: $\rho = 0$; that is, no linear relationship exists between **X** and **Y**. One may equivalently test H_0: **B** $= 0$, for if there is no linear relationship, then ρ must be zero. However, there are separate statistical significance tests for $\rho = 0$ (see References 1 and 13, for example).

Naturally, a calculated **r** would seldom be exactly zero in such a situation. The statistical test specifies critical values for **r** that are so far from zero that they would occur with low (.05 or .01) probability under H_0. If our calculated **r** is found to be not statistically significant (**p** $> .05$), we know that the probability of obtaining such a value (or one even further from zero) is at least

_____.

.05

COEFFICIENT OF DETERMINATION

6.24 Another statistic that is useful in the study of the linear relationship of measurements is the **coefficient of determination**, r^2 **or** R^2, which is merely the square of the calculated correlation coefficient. Obviously, the range of r^2 is _____ $\leq r^2 \leq$ _____.

0, 1

6.25 The r^2 ($\times$ 100%) statistic measures the proportion (%) of the **Y** sums of squares $\Sigma(Y - \overline{Y})^2$ that may be "explained" by the regression formula $Y = a + bX$. That is, if all the points on a scattergram lie on a straight line,

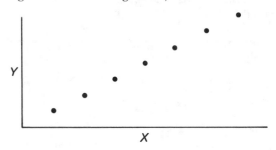

all the variability as measured by the **Y** sums of squares may be accounted for (or explained by) knowing **X**. Given a value of **X**, the corresponding **Y** value is known without error if $r = +1$ or -1, thus yielding a coefficient of determination of _____ .

1

6.26 Therefore, if we have

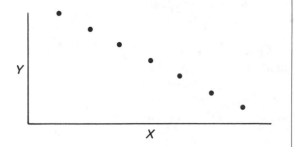

then $r =$ _____ and $r^2 =$ _____ . The proportion of the **Y** sums of squares that may be accounted for by this linear regression is ___%.

$-1, 1$

100

6.27 If $r = 0$, the proportion of **Y** sums of squares that may be accounted for by regression is zero. Other values of **r** yield $0 < r^2 \le 1$, which provides a more useful **measure of the predictability** of **Y** from **X** than the correlation coefficient alone.

The coefficient of determination is a meaningful statistic to calculate in both experimental and nonexperimental studies. It should always be considered when interpreting correlation coefficients, because a correlation coefficient may be statistically significant, yet small. Hence, the coefficient of determination will be small. This situation is especially likely to occur with large sample sizes.

If $n = 100$ and the calculated r turns out to be $-.30$, for example, this is statistically significant at the .01 level ($p < .01$). Therefore, we are fairly certain that $\rho \neq 0$ (i.e., that there is a linear relationship between X and Y). However, we can explain only _____% of the Y sums of squares | 9
by regression on X. Therefore, the predictability of Y from X is _____ (good, poor). | poor

6.28 The concept of linear regression may be extended from one independent variable to two or more. Thus, we may write $Y = A + B_1X_1 + B_2X_2 + \ldots + B_KX_K$, where A and the Bs are coefficients and the Xs are variables. Such a formula is called a **multiple linear regression equation**.

For example, it may be useful to write Y (probability of death in the next 5 years) $= A + B_1X_1$ (age) $+ B_2X_2$ (average systolic blood pressure) $+ B_3X_3$ (average serum cholesterol level) $+ B_4X_4$ (a measure of obesity) $+ \ldots$ (and so forth).

The coefficient of determination may also be calculated for multiple linear regression situations. Although the formula is different, the interpretation is the same as that of simple (one X) linear regression; that is, R^2 (or r^2) is the proportion of the Y sums of squares that may be _____ by the multiple linear regression. The | explained (or
computation of R^2 is useful in assessing the pre- | accounted for)
dictability or utility of such equations.

Clinical Trials

7

EXPERIMENTS

On the basis of the types of inferences that may be drawn, scientific studies may be divided into **experiments** and **nonexperimental studies**. The latter, which include population surveys, are sometimes referred to as **observational studies**. "Observational studies" seems a poor term, however, since it apparently implies that observations are not made in experiments.

The difference is that in an experiment, the investigator has control over the major "independent" or "predictor" variable(s) (i.e., the drugs, treatments, etc.) under study and **assigns their values** to the experimental units (persons, animals, cell cultures, and the like). In a nonexperimental study, there is no such control, and the investigator studies the relationships among variables **as they occur in nature**.

In nonexperimental studies, researchers look for associations among variables, often in a search for possible cause-and-effect relationships. Because of the complex relationships that may exist among both observed and unobserved variables, however, it is nearly always inappropriate to assume that an association among variables implies that they are causally related. **Such inference is proper only in experiments**, where it is possible to rule out other competing hypotheses, at least for most situations encountered in biomedical research.

The fundamental elements of experimentation may be summarized as follows:

1. Specification of the objectives. This usually requires stating the hypothesis to be tested.
2. Specification of the experimental unit (animals, humans, cells, etc.).
3. Specification of the appropriate experimental design, including the choice of the control or comparison group(s).
4. Random allocation of experimental units to study (e.g., treatment) groups.

5. Making every reasonable effort to ensure that study groups are treated as exactly alike as possible throughout the course of the experiment.

7.1   Items 4 and 5 above are essential to ensure that any apparent treatment differences are due only to (1) true treatment differences or (2) random processes.

A study in which morning patients received treatment A and afternoon patients received treatment B, for example, _____ (would, would not) constitute a random assignment of subjects (experimental units) to treatment groups.

would not

CLINICAL TRIALS

7.2   A **clinical trial** is an experiment in which the experimental unit is a human. The same general principles of good experimentation apply, of course, but there are some areas of special interest.

Until a few years ago, most clinical research was based upon nonexperimental (sometimes called "uncontrolled") studies. However, there has been a striking change to emphasize experiments: the introduction of clinical trials. Today, most clinical researchers would prefer clinical trials in the evaluation of new surgery, drugs, or other therapies whenever possible. Indeed, United States Food and Drug Administration policies require sound clinical trial results before a new drug is approved. We will next examine some of the features of this research method.

The major scientific advantage of clinical trials over nonexperimental studies of therapeutic procedures is that _____ inferences may be drawn.

causal

7.3   Because of the highly variable nature of responses encountered in clinical research, nearly all clinical experiments are comparative. Seldom is the variability so small that the use of "historical" controls (i.e., in which new results are compared with past rates or values) would be acceptable. Therefore, simultaneous comparison of treatments is usually _____ (desirable, undesirable).

desirable

7.4 Two or more therapeutic treatments usually characterize the treatment groups to be studied. In drug investigations, one treatment often is a "control" or **placebo**. An ideal placebo is a biologically inert substance that physically resembles the active drug. It is well known that placebo effects (presumably psychologic) can be substantial in humans. Therefore, such a control group is usually _____ (required, not required) in drug trials.

required

7.5 The experimental requirement (item 5 above) of keeping "everything else constant" often presents special problems in clinical trials. It is usually much more difficult to control the environmental factors throughout such a study than it is in experiments that do not utilize persons as experimental units.

Psychologic variables can be particularly troublesome. For example, volunteer subjects participating in a trial **who know** which medication they are receiving may allow this to influence their own assessment of their well-being. If part of the treatment evaluation is their own subjective evaluation of their condition, this knowledge would bias the results. Random assignments of subjects to treatment groups _____ (would, would not) remove this bias.

would not

7.6 Therefore, if the experiment is a drug trial, it is often possible to code the material so that the subject does not know which of the treatments he or she is receiving; this is called a **single-blind clinical trial**. A **double-blind clinical trial** is one in which neither the subjects nor the investigators know which treatments are being given to specific subjects until the end of the study when the code is broken and the data are analyzed. Of course, single-blind and double-blind trials are not always feasible, but when they are, such procedures tend to _____ (minimize, maximize) the bias that could be caused by subjective evaluations.

| | minimize |

7.7 A _____ clinical trial is one in which neither the subjects nor the investigators know which treatments are being given to individual patients until after all the data are recorded.

| | double-blind |

7.8 Of course, all persons participating as subjects in a clinical trial must be volunteers. "Informed consent" must be obtained before a human may become part of a clinical experiment. If volunteers differ from nonvolunteers with respect to their response to treatment in some major way, then inferences _____ (may, may not) legitimately be extended to the entire population of interest. This constraint is one that must be considered in the interpretation of all trials.

| | may not |

7.9 Volunteers are randomly assigned or allocated to treatment groups. According to our previous understanding of randomness, this means that each subject has a(n) _____ chance of being assigned to each treatment.

| | equal |

7.10 Clinical trial results may be assessed statistically like any other set of experimental data. Statistical tests are used to judge objectively whether observed treatment differences are likely due to true treatment differences or to chance (see section 7.1). Because the relative degree of variation in measurements made on humans in experimental and nonexperimental studies is greater than that encountered in most other sciences, clinical researchers often depend more heavily on statistical methods to assess the role of randomness or chance in their observations. For this reason clinical trials and epidemiology are singled out for emphasis in this book.

Suppose 20 volunteers in a double-blind clinical trial are randomly allocated to two treatments and are observed for 1 month. Simple improvement rates are found to be as follows:

Treatment	Improved	Unim-proved	Rate
A	15	5	15/20 = 75%
B (placebo)	10	10	10/20 = 50%

An appropriate statistical test that could be employed here is a _____ (χ^2, t) test.

χ^2

7.11 If a similar trial employed measurements in the evaluation instead of rates, then a _____ (χ^2, t) test would be more appropriate.

t

Obviously, we have only touched on the most basic principles of clinical trials. There are many types of such experimental designs. Special clinical, ethical, inferential, and statistical problems emerge as medical science continues to refine this important method. As with any successful scientific method or model, this one has its limitations. Discussion of these issues may be found in several texts (including References 1, 12, and 15).

One important distinction is that between **therapeutic trials** (comparing methods of treating diseased patients) and **prophylactic or prevention trials** (comparing methods of preventing disease). Both are experimental, but different ethical and scientific issues can arise in each situation. (See References 5, 6, and 12.)

Epidemiology

8

EPIDEMIOLOGY AND CAUSAL INFERENCE

Epidemiology is the study of the distribution and determinants of health-related states or events in specified populations, and the application of this study to control of health problems. While there have been many definitions of this discipline, the preceding, which appears in Last's **A Dictionary of Epidemiology** (Reference 10), represents as close to a consensus as any available. A key point is the population-based perspective.

8.1 An epidemiologist often attempts to characterize a disease by studying variation in prevalence or incidence as it relates to time, geography, and many population characteristics. By necessity, such studies nearly always are of a disease as it naturally occurs. Thus, epidemiology _____ (is, is not) primarily an experimental science. Because of this frequent restriction (lack of experimental data in humans), epidemiologists are often forced to make inferences about disease causation based mainly on nonexperimental evidence.

is not

As Friedman (Reference 5) states, "Experiments are believed to be the best test of a cause-and-effect relationship." Because it is usually not possible to conduct such studies of suspected disease causes in human populations (for ethical and logistic reasons), careful examination of **observed associations** based on specified criteria has been suggested. One such set was formulated by Hill (Reference 7). These deal with the **strength, consistency** (study-to-study), and **specificity** of the association. Other factors are the **temporal logic** (suspected cause must precede the effect), **biologic gradient** (is there a dose-response curve?), and **biologic plausibility**. Overall judgments in public health issues utilize the total coherence of the evidence based on these criteria incorporating epidemiologic and all other relevant information.

2.2 Although rare, a clinical trial of a prophy-
actic measure is considered to provide the most
definitive evidence. One example is the complex
and long-term double-blind evaluation of the effects of drug-induced serum cholesterol lowering
on the occurrence of coronary heart disease in
men with high cholesterol levels (Reference 11).
This trial did demonstrate such an effect. Because
it was a(n) _____ (experiment, nonexperimental study) with high power, competing causal hypotheses could be ruled out with reasonable probability.

experiment

NOTE: Once a sufficient scientific or societal consensus is reached regarding causality, public
health action is usually implemented. This process may take years, however, as such agreement
is not quickly obtained, especially when various
studies of the same issue are in conflict. It is instructive to consider some recent examples.

Causal inference is a complex topic that un-
derlies the basis of all science. Presented here are
the barest fundamentals that have been accepted
by most researchers in the field. However, the
subject is a dynamic one with alternative views
and ideas continually being expressed. See References 5, 6, 7, 8, 9, 14, and 15 for further discussion.

STUDY DESIGNS IN EPIDEMIOLOGY

2.3 An epidemiologist is interested in that which
differentiates the sick (usually from a specific illness) from the well in a defined population. The
basic investigative tools available to the epidemiologist for examining this question are the **cohort** (or prospective) **study** and the **case-control**
(or retrospective) **study**.

A **cohort** epidemiologic study is one in which
it is determined for a number of persons whether
or not they have been exposed to a suspected

causal variable, **X**. That is, it is suspected that exposure to **X** could contribute to the development of a disease. To examine this hypothesis, at a specific later point in time, disease incidence rates are usually compared.

This is illustrated in the simplest type of design with two exposure groups, E (exposed) and $\bar{E}$ (unexposed), and two outcome groups, D (diseased) and $\bar{D}$ (no disease). The typical resulting 2×2 table is shown below.

Number of persons

| Exposure status | Disease status | | |
	D	$\bar{D}$	Total
E	a	b	a+b
$\bar{E}$	c	d	c+d
Total	a+c	b+d	n

It would be reasonable to compare disease incidence rates in the two exposure groups; that is,

$$\frac{a}{a+b} \text{ versus } \underline{\hspace{2cm}}$$

$\dfrac{c}{c+d}$

8.4 Another term for the exposed (E) incidence rate $\dfrac{a}{a+b}$ is the **risk** of acquiring the disease in the specified time period of the cohort "follow-up" period. The risk of disease in the _____

unexposed ($\bar{E}$)

(exposed, unexposed) group is $\dfrac{c}{c+d}$.

The ratio of these two risks is called the **relative risk (RR)**. It simply compares disease risk for the suspected risk factor in the exposed to the risk in the unexposed. Thus, the ratio $\dfrac{a/a+b}{c/c+d}$ is an estimate of the _____.

relative risk (RR)

8.5 In the following cohort study example, what is the estimated relative risk? _____

$RR = \dfrac{40/100}{20/200} = 4.0$

| Risk factor | Number of persons | | |
| status | Disease status | | |
	D	$\bar{D}$	Total
E	40	60	100
$\bar{E}$	20	180	200
Total	60	240	300

8.6 If the preceding sample data are close to representing the population to which the results are to be generalized, the risk of a person's contracting the disease (during the period of the study) when exposed to the risk factor is about _____ times that for those not exposed. Confidence intervals can and should be calculated to accompany any measures of relative risk. A description of methods applicable in many situations is found in Reference 6.

four

8.7 Suppose an **RR** calculation turned out to be exactly 1.0. The conclusion of the study would be that the risk factor was associated with _____ (an increase, a decrease, no change) in the disease rate. If the **RR** was estimated to be 0.5, the risk factor would be associated with a disease _____ (increase, decrease).

no change
The risks were identical in the two groups.
decrease
The evidence supports a "protective" effect.

8.8 The **RR** is the key summary statistic provided in most cohort epidemiologic studies. However, in the overall assessment from the population perspective, the absolute risk must also be taken into account. Two incidence rates of 2 per 100,000 (E) and 1 per 100,000 ($\bar{E}$) yield a relative risk of _____. However, even if real, the difference in absolute risk (1 per 100,000) may be practically negligible. Both the relative risk and the absolute risk (with a defined time period) should be borne in mind when evaluating disease risks.

2.0

8.9 Cohort studies can be expensive to conduct, especially if the disease under study is rare. One might have to follow thousands of people for several years to observe a sufficient number of cases of disease that would provide reasonable estimates of absolute and relative risk. If only a few cases are observed, the statistical power of a study is likely to be _____ (low, high).

low

8.10 A less costly alternative approach, especially in the study of chronic diseases that develop over many years, is the **case-control** or case-comparison method. In this type of investigation, the number of persons in a defined population who have a particular disease is noted. A comparison ("control") group is then selected from the population. (Choosing such a group appropriately is an exceedingly challenging task.) It is then determined (as well as possible) how many (or to what degree) persons in each of the two groups had been exposed to the suspected causal or exposure variable. Now, it is not the _____ (exposure, incidence) rates that are compared, as in the cohort study; it is more logical to compare the _____ (exposure, incidence) rates in the diseased and well groups.

incidence

exposure

8.11 Consider the 2×2 table that would be formed by the case-control method. It appears identical with that obtained by the cohort methods shown in Section 8.3.

Number of persons

Exposure status	Disease status		
	D	$\overline{D}$	Total
E	a	b	a + b
$\overline{E}$	c	d	c + d
Total	a + c	b + d	n

In this case, however, the table was formed by starting with the groups D and $\overline{D}$ and noting the number of exposed and unexposed. The two exposure rates in this case are $\dfrac{a}{a+c}$ and _____.

$\dfrac{b}{b+d}$

8.12 An apparent food-poisoning outbreak may be used to illustrate the case-control approach. A food consumption history for both sick and well persons would be obtained, and a 2×2 table for each food item (suspected causal variable) would be calculated. Divergent exposure rates would implicate certain foods as the source of the food poisoning. Naturally, some variation between these rates would occur by chance, even if there is no real difference. The appropriate statistical test to use here would be the _____ test.

$2 \times 2 \; \chi^2$

8.13 An example of such an epidemiologic survey of a food-poisoning outbreak involving 70 persons is summarized as follows:

Salad

	Ate	*Did not eat*	Total
Sick	50	0	50
Not sick	20	0	20

Soup

	Ate	*Did not eat*	Total
Sick	40	10	50
Not sick	5	15	20

Ice cream

	Ate	*Did not eat*	Total
Sick	25	25	50
Not sick	10	10	20

The food most strongly implicated epidemiologically is _____.

soup (40/50 versus 5/20, or 80% versus 25%)

8.14 As stated, case-control studies of chronic diseases may also be conducted. A well-known example is the comparison of cigarette exposure (**X**) rates in persons with and without lung cancer. A major drawback in chronic disease studies, however, is the length of time (usually years) between the exposure to a suspected causal variable and the manifestation of the disease. If exposure to **X** affects the probability of a person's remaining in the population until the time of termination of the study, then the calculated exposure rates are clearly _____ (biased, unbiased).

biased

8.15 The _____ (longer, shorter) the time period between the time of exposure to a suspected causal variable and the time of the investigation, the smaller the likelihood of obtaining accurate exposure histories. Useful relationships may still be uncovered, however, even if years have elapsed. For example, rigorous epidemiologic (case-control) investigation revealed the link between ingestion of diethylstilbestrol by pregnant women and the occurrence of adenocarcinoma of the vagina in daughters 18 to 25 years later.

longer

8.16 In chronic disease case-control studies in which the latency period (approximate time from exposure to resulting disease) is a matter of years, another important measure of association is often used, the **odds ratio**. Consider, again, the general 2×2 table for case-control studies shown in Section 8.11. We may say the "odds" of getting the disease, relative to not getting the disease, for the exposed group is **a/b**. The odds of getting the disease, relative to not getting the disease, for the unexposed group is **c/d**. The ratio of these two odds is the **odds ratio**

$$\frac{a/b}{c/d}$$

which algebraically reduces to $\dfrac{\mathbf{ad}}{}$, a simple ratio of cross-products in the 2×2 table.

$$\frac{\mathbf{ad}}{\mathbf{bc}}$$

8.17 While perhaps not as intuitively appealing as the relative risk measure used in _____ studies, the odds ratio has become a commonly calculated statistic in _____ studies.

cohort

case-control

8.18 It turns out that (given that dropout and other biases are minimal) the odds ratio is approximately equal to the relative risk estimate in situations in which the disease being studied is of low prevalence. This can be seen by examining the identical summary 2×2 tables of sections 8.3 and 8.11. Recall that the relative risk estimate is $\dfrac{a/(a+b)}{c/(c+d)}$. When **a** is very small compared to **b** and **c** is very small compared to **d** (the case with a low prevalence disease), then we may write the approximation

$$\frac{a/(a+b)}{c/(c+d)} \approx \frac{a/b}{c/d} = \frac{ad}{bc}$$

which is equivalent to the formula for the _____ _____. Thus, we sometimes see the two terms **relative risk** and **odds ratio** used interchangeably in studies of disease with _____ (high, low) prevalence.

odds

ratio

low

8.19 Cohort studies are less susceptible than case-control studies to unknown influences on the population. Such epidemiologic surveys — provided the comparison-control group is carefully chosen — more closely resemble the experiment that would be required to establish a causal factor. One major disadvantage of cohort studies of most diseases is that the incidence rate for any defined "exposure group" is quite low; therefore, extremely large sample sizes may be required in order to observe enough persons with the disease for the study to be meaningful. Small sample sizes may mean the study lacks _____.

power

In a cohort study in which the exposed and unexposed groups differ by no discernible characteristics except their exposure to **X**, a large difference in _____ (exposure, incidence) rates

incidence

still does not prove that this variable, **X**, is part of the causal mechanism in the production of the disease, because the "assignment" of **X** to the two groups was not random. The investigator did not allocate the "treatment" **X** to the study groups (i.e., natural or self-selection is involved to an indeterminable extent). An association detected in this manner may indicate the need for further study to determine the causative role, if any, of this factor in the population. Of course, in most situations, ethical considerations preclude conducting the definitive experiment to establish disease causality.

8.20 A simple general population survey can provide useful data for epidemiologists. A **cross-sectional** (or **prevalence**) **study** allows examination of the relationship between disease occurrence and other variables as they exist at a particular point in time. If a 2×2 table is constructed, the relationship may be analyzed either by comparing attack rates or by comparing _____ rates.

exposure

The National Center for Health Statistics periodically conducts such surveys; one, the annual Health Interview Survey, is based upon perceived health status and another, the National Health and Nutrition Examination Survey, utilizes a wide range of medical, dental, psychologic, and other examinations. These provide valuable, descriptive, cross-sectional information on the noninstitutionalized United States adult population.

STATISTICAL ANALYSIS IN EPIDEMIOLOGIC STUDIES

8.21 In uncomplicated studies like those just presented involving the comparison of two groups, simple statistical tests, such as the _____ test for comparing rates or the _____ test for comparing the means of two sets of measurements, are often adequate for assessing the statistical significance of an observed association.

$2 \times 2 \, \chi^2$, **t**

The power of such tests should be kept in mind, however, in interpreting nonsignificant results because small sample sizes imply _____ (high, low) power, which may mean that the probability of detecting an important group difference (association between suspected cause and disease) is unacceptably _____ (high, low).

low

low

8.22 Variables (discrete or continuous) suspected of causally contributing to the development of a disease are called **risk factors** for that disease. In epidemiologic studies designed to first assess whether or not such variables are related to the disease under study, sample means or percentages are usually compared by statistical tests. When both a suspected risk factor and the disease are measurement (continuous) variables a _____ _____ is often calculated and tested for statistical significance. A scatter diagram must also be plotted, as discussed in Chapter 6. All such tests assess whether or not there is an association between the suspected risk factor and the disease. Of course that is only a first step; association alone does not imply causation.

correlation
coefficient

8.23 In addition to such tests for association, the **RR** or **relative risk** (in cohort studies) and the **odds ratio** in (_____-_____) studies are calculated as valuable summary measures. If there is in reality no association between exposure to the suspected risk factor and disease then the long-run average **RR** and odds ratio would be _____ (zero, one) in repeated studies from the same populations. Therefore, it is clearly of interest to examine the confidence intervals of these statistics to see if they include the number one. Inclusion of one in a such 95% confidence intervals is equivalent to accepting the null hypothesis, H_0, of no association, at the 5% level of significance. Conversely, if such an interval excludes one, we can _____ (accept, reject) H_0 at the 5% level.

case-control

one

reject

8.24 The examples presented here are an oversimplification of epidemiologic studies, especially in chronic disease situations, in which the **latency period** may be years. Of course, exposure may not be at a single point in time. This time factor can allow many covariables or cofactors to play a role. If another variable occurs in such a way as to distort or bias the apparent association, it is called a **confounding** variable. Such confounding can occur in experiments as well but is much more prominent (and frustrating) in nonexperimental studies. If, in a study of cigarette smoking and disease occurrence, it is determined that all cigarette smokers are also heavy consumers of alcohol, those two variables are said to be _____.

confounded

NOTE: The most difficult scientific aspect of epidemiologic research is the selection of the control or comparison groups. Innumerable biases (known and unknown) can frustrate the researcher in the attempts to minimize bias and deal with confounding.

Chronic diseases usually have multiple etiologies and involve several (known and unknown) variables. In addition, some covariables may not be regarded as causal but may be "promoters." Some may be related to the causal variables but incidental to the disease. Such complexities need to be taken into account in the design, analysis, and interpretation of most epidemiologic studies.

It is almost always logistically and economically infeasible to conduct a study large enough to properly analyze all combinations of relevant variables separately. **Multivariate statistical methods** have been derived to model such relationships in a reasonable manner. Models such as **logistic regression** and **Cox** (or professional hazards) **regression** have proven eminently useful in such cases. It is often possible to adequately estimate relative risks with such techniques, even with a large number of covariables. No amount of modeling or analysis can truly adjust for serious confoundings, however. See References 5, 8, and 9 for discussion of multivariate analyses in epidemiology and their interpretation.

References

1. Colton, T. *Statistics in Medicine*. Boston: Little, Brown, 1974.
2. Duncan, R. C., Knapp, R. G., and Miller, M. C., III. *Introductory Biostatistics for the Health Sciences* (2nd ed.). New York: Wiley, 1983.
3. Elveback, L. R., et al. Health, normality, and the ghost of Gauss. *J.A.M.A.* 211:69, 1970.
4. Fleiss, J. L. *Statistical Methods for Rates and Proportions* (2nd ed.). New York: Wiley, 1981.
5. Friedman, G. D. *Primer of Epidemiology* (3rd ed.). New York: McGraw-Hill, 1986.
6. Hennekens, C. H., and Buring, J. E. *Epidemiology in Medicine*. Boston: Little, Brown, 1987.
7. Hill, A. B. The environment and disease: Association or causation? *Proc. R. Soc. Med.* 58:295, 1954.
8. Kahn, H. A., and Sempos, C. T. *Statistical Methods in Epidemiology*, 2 ed. New York: Oxford, 1989.
9. Kleinbaum, D. G., Kupper, L. L., and Morgenstern, H. *Epidemiologic Research*. Belmont, CA: Life-time Learning, 1982.
10. Last, J. M. (ed.). *A Dictionary of Epidemiology* (2nd ed.). New York: Oxford, 1988.
11. Lipid Research Clinics Program. The Lipid Research Clinics Coronary Primary Prevention Trial Results. I. Reduction in incidence of coronary heart disease. *J.A.M.A.* 251:351, 1984.
12. Meinert, C. L. *Clinical Trials*. New York: Oxford, 1986.
13. Rosner, B. *Fundamentals of Biostatistics* (3rd ed.). Boston: PWS–Kent, 1990.
14. Rothman, K. J. Causation and Causal Inference. In D. Schottenfeld and J. F. Fraumini (eds.), *Cancer Epidemiology and Prevention*. Philadelphia: Saunders, 1982.
15. Sackett, D. L., Haynes, R. B., and Tugwell, P. *Clinical Epidemiology*. Boston: Little, Brown, 1985.

Index